FOURIER
SERIES
AND
BOUNDARY
VALUE
PROBLEMS

RUEL V. CHURCHILL

PROFESSOR OF MATHEMATICS
UNIVERSITY OF MICHIGAN

McGRAW-HILL BOOK COMPANY

NEW YORK   SAN FRANCISCO   TORONTO   LONDON

FOURIER SERIES AND BOUNDARY VALUE PROBLEMS

SECOND EDITION

FOURIER SERIES AND BOUNDARY VALUE PROBLEMS

4 5 6 7 8 9 – M P – 70 9 8 7 6

10841

PREFACE

This is an introductory treatment of Fourier series and their applications to boundary value problems in partial differential equations of engineering and physics. It is designed for students who have completed the equivalent of one semester of advanced calculus. The physical applications, explained in some detail, are kept on a fairly elementary level.

The first objective is to introduce the concept of orthogonal sets of functions and representations of arbitrary functions in series of the functions of such sets. The most prominent special cases, the representation of functions by trigonometric Fourier series, are given special attention. Fourier integral representations and expansions in series of Bessel functions and Legendre polynomials are also treated.

The second objective is a clear presentation of the classical method of solving boundary value problems with the aid of those representations in series of orthogonal functions. Some attention is given to the verification of solutions and to uniqueness of solutions, for the method cannot be presented properly without such considerations. Other methods are treated in the author's books "Operational Mathematics" and "Complex Variables and Applications."

This edition is an extensive revision of the original 1941 edition of the book. The exposition has been revised throughout. Some additional material has been introduced on differential equations and boundary conditions, uniform convergence, complex-valued functions, Fourier integrals, convergence of Legendre's series, uniqueness of solutions, and other topics. Some rearrangement of topics was found desirable; for instance, partial differential equations of physics are now treated in the first chapter in order to simplify the introduction of other topics.

Additional attention is given to the mathematical analysis.

Examples, problems, figures, and bibliography have been revised.

The chapters on Bessel functions and Legendre polynomials, Chapters 8 and 9, are independent of each other. They can be taken up in either order. Chapter 10, on uniqueness of solutions, and Chapter 5, on further properties of Fourier series, as well as some sections of other chapters, can be omitted in order to shorten the course.

In the development of the book through this edition the author acknowledges the helpful comments and encouragement of many teachers and students. Among his local colleagues, Professors R. C. F. Bartels, C. L. Dolph, G. E. Hay, and E. D. Rainville deserve special thanks.

RUEL V. CHURCHILL

CONTENTS

CHAPTER 1

PARTIAL DIFFERENTIAL EQUATIONS OF PHYSICS

1. Two Related Problems. We shall be concerned here with two general types of problems. One type deals with the representation of arbitrarily given functions by infinite series of functions of a prescribed set. The other consists of boundary value problems in partial differential equations, with emphasis on equations that are prominent in physics and engineering.

Representations by series are encountered in methods of solving boundary value problems. The theories of those representations can be presented independently. They have such attractive features as the extension of concepts of geometry, vector analysis, and algebra into the field of mathematical analysis. Their mathematical precision is also pleasing. But they gain in unity and interest when presented in connection with boundary value problems.

The set of functions that make up the terms in the series representation is determined by the boundary value problem. Representations or expansions in Fourier series, certain types of series of sine or cosine functions, are associated with the more common boundary value problems. We shall give special attention to the theory and application of Fourier series. But we shall also consider extensions and generalizations of such series, including Fourier integrals and series of Bessel functions and Legendre polynomials.

A boundary value problem is correctly set if it has one and only one solution. Physical problems associated with partial differential equations often suggest boundary conditions under which a problem may be correctly set. In fact, it is sometimes helpful to interpret a problem physically in order to judge whether the boundary conditions may be adequate. This is a prominent reason for associating such problems with their physical applications, aside from the opportunity to display interesting and important contacts between mathematical analysis and the physical sciences.

The theory of partial differential equations gives results on the existence of solutions of boundary value problems. But such results are necessarily limited and complicated by the great variety of features: types of equations and conditions, and types of domains. Instead of appealing to general theory in treating a specific problem, we may actually find a solution and then prove that only one solution is possible.

2. Linear Boundary Value Problems. Theory and applications of ordinary or partial differential equations in a function u usually require that u satisfies not only the differential equation throughout some domain of its independent variable or variables but also some conditions on boundaries of that domain. The equations that represent those boundary conditions may involve values of derivatives of u, as well as u itself, at points on the boundary. In addition, some conditions on the continuity of u and its derivatives within the domain and at the boundaries are required.

Such a set of requirements constitutes a *boundary value problem* in the function u. We apply that term whenever the differential equation is accompanied by some boundary conditions even though the conditions may not be adequate to ensure a unique solution of the problem.

The three equations

$$(1) \qquad \begin{aligned} u''(x) - u(x) &= -1 \qquad (0 < x < 1), \\ u'(0) &= 0, \qquad u(1) = 0, \end{aligned}$$

for example, constitute a boundary value problem in ordinary differential equations. The domain of the independent variable x is the interval $0 < x < 1$ whose boundaries consist of the two points $x = 0$ and $x = 1$. The solution of this problem which, together with each of its derivatives, is continuous everywhere is found to be

$$(2) \qquad u(x) = 1 - (\cosh 1)^{-1} \cosh x.$$

Frequently it is convenient to indicate partial differentiation by writing independent variables as subscripts. If, for instance,

u is a function of the independent variables x and y, we may write

$$u_x \text{ or } u_x(x,y) \text{ for } \frac{\partial u}{\partial x}, \qquad u_{xx} \text{ for } \frac{\partial^2 u}{\partial x^2}, \qquad u_{xy} \text{ for } \frac{\partial^2 u}{\partial y\, \partial x},$$

and so on. Also, we shall be free to use the symbols $u_x(x_0,y)$ and $u_{xx}(x_0,y)$ to denote the values of the functions $\partial u/\partial x$ and $\partial^2 u/\partial x^2$, respectively, on the line $x = x_0$ and corresponding symbols for boundary values of other derivatives.

The problem consisting of the partial differential equation

$$(3) \qquad u_{xx}(x,y) + u_{yy}(x,y) = 0 \qquad (x > 0,\, y > 0)$$

and the two boundary conditions

$$(4) \qquad \begin{aligned} u(0,y) &= u_x(0,y) & (y > 0),\\ u(x,0) &= \sin x + \cos x & (x > 0), \end{aligned}$$

is an example of a boundary value problem in partial differential equations. The domain is the first quadrant of the xy plane. The reader can verify that the function

$$(5) \qquad u(x,y) = e^{-y}(\sin x + \cos x)$$

is a solution of that problem. This function and its partial derivatives are everywhere continuous in the two variables x,y together and bounded in the domain $x > 0$, $y > 0$.

A differential equation in a function u, or a boundary condition on u, is *linear* if it is an equation of the *first degree in u and derivatives of u.* Thus the terms of the equation are either functions of the independent variables alone, including constants, or such functions multiplied by either u or one of the derivatives of u.

The differential equations and boundary conditions (1), (3), and (4) above are all linear. The differential equation

$$(6) \qquad z u_{xx} + x y^2 u_{yy} - e^x u_z = f(y,z)$$

in $u(x,y,z)$ is linear. But the equation

$$u_{xx} + u u_y = x$$

is nonlinear in $u(x,y)$ because the term $u u_y$ is not of the first degree as an algebraic expression in the two variables u and u_y.

Let the letters A to G denote either constants or functions of the independent variables x and y only. Then the general *linear* *partial differential equation of second order in $u(x,y)$* has the form

$$(7) \qquad Au_{xx} + Bu_{xy} + Cu_{yy} + Du_x + Eu_y + Fu = G.$$

A boundary value problem is *linear* if its differential equation and all its boundary conditions are linear. Problem (1) and the problem consisting of equations (3) and (4) are examples of linear boundary value problems.

Methods of solution presented in this book do not apply to nonlinear boundary value problems.

A linear differential equation or boundary condition is *homogeneous* if each of its terms, other than zero itself, is of the first degree in the function u and its derivatives.

Equation (7) is homogeneous in a domain if and only if $G(x,y) = 0$ throughout that domain. Equation (6) is nonhomogeneous if $f(y,z) \neq 0$. Equation (3) and the first of conditions (4) are homogeneous. In our treatment of linear boundary value problems, homogeneous equations will play a distinctive role.

3. The Vibrating String. A tightly stretched string, whose position of equilibrium is some interval on the x axis, is vibrating in the xy plane. Each point of the string, with coordinates $(x,0)$ in the equilibrium position, has a transverse displacement $y(x,t)$ at time t. We assume that the displacements y are small relative to the length of the string, that slopes are small, and that other conditions are such that the movement of each point is essentially in the direction of the y axis. Then at time t the point has coordinates (x,y).

Let the tension P of the string be great enough that the string behaves as if it were perfectly flexible; that is, at each point the part of the string on the left of that point exerts the force of magnitude P in the tangential direction upon the part on the right; the effect of bending moments at the point can be neglected. The magnitude of the x component of the tensile force is denoted by H (Fig. 1). Our final assumption is that H is constant, that is, that the variation of H with x and t can be neglected.

Those idealizing assumptions are severe; but they are justified in many applications. They are adequately satisfied, for instance, by strings of musical instruments under ordinary conditions of operation. Mathematically, the assumptions lead to a partial differential equation in $y(x,t)$ which is linear.

Now let $V(x,t)$ denote the y component of the tensile force exerted by the left-hand portion of the string on the right-hand portion at the point (x,y). We take the positive sense of V as that of the y axis. If α is the slope angle of the string at the point (x,y) at time t, then $-V/H = \tan\alpha = \partial y/\partial x$ as indicated in Fig. 1. Thus *the y component V of the force exerted by the part of the string on the left of a point (x,y) upon the part on the right, at time t,* is given by the formula

$$(1) \qquad\qquad V(x,t) = -Hy_x(x,t) \qquad\qquad (H > 0).$$

This is the basic formula for deriving the equation of motion of the string. It is also used in setting up certain types of boundary conditions.

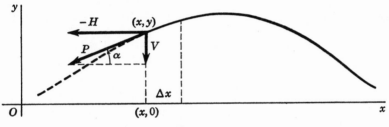

Fig. 1

Suppose that all external forces such as the weight of the string and resistance forces, which act on the string, other than forces at the end points, can be neglected. Consider a segment of the string not containing an end point, whose projection on the x axis has length Δx. Since x components of displacements are negligible, the mass of the segment is $\delta\,\Delta x$, where the constant δ is the mass of the string per unit length. At time t the y component of the force exerted by the string on the segment at the left-hand end (x,y) is $V(x,t)$, given by formula (1). The y component of the force exerted by the string on the other end of the segment is $-V(x + \Delta x, t)$, where the negative sign signifies that the force is exerted by the right-hand part upon the left-hand part at that point. The acceleration of the end (x,y) in the y direction is $y_{tt}(x,t)$. According to Newton's second law of motion (mass times acceleration equals force), then

$$(2) \qquad \delta\,\Delta x\, y_{tt}(x,t) = -Hy_x(x,t) + Hy_x(x + \Delta x, t),$$

approximately, when Δx is small.   Hence

$$y_{tt}(x,t) = \frac{H}{\delta} \lim_{\Delta x \to 0} \frac{y_x(x + \Delta x, t) - y_x(x,t)}{\Delta x} = \frac{H}{\delta} y_{xx}(x,t)$$

at each point where the partial derivatives exist.

Thus the function $y(x,t)$, representing the transverse displacements in a stretched string under the conditions stated above, satisfies the *wave equation*

(3)
$$\frac{\partial^2 y}{\partial t^2} = a^2 \frac{\partial^2 y}{\partial x^2} \qquad \left(a^2 = \frac{H}{\delta} > 0\right)$$

at points where no external forces act on the string.   The constant a has the physical dimensions of velocity.

4. Modifications of the Equation. End Conditions.   When external forces parallel to the y axis act along the string, let F denote the force per unit length of string.   Then a term $F \Delta x$ must be added to the right-hand member of equation (2), Sec. 3, and the equation of motion is

(1)
$$y_{tt}(x,t) = a^2 y_{xx}(x,t) + F\delta^{-1}.$$

In particular, if the y axis is vertical with its positive sense upward and the external force consists of the weight of the string, then $F \Delta x = -\delta \Delta x\, g$, where the constant g is the acceleration of gravity.   Equation (1) then becomes the linear nonhomogeneous equation

(2)
$$y_{tt}(x,t) = a^2 y_{xx}(x,t) - g.$$

In equation (1), F may be a function of x, t, y, or derivatives of y.   In case the external force per unit length is a damping force proportional to the velocity in the y direction, for example, F is replaced by $-By_t$, where the positive constant B is a coefficient damping.   Then the equation of motion is linear homogeneous:

(3)
$$y_{tt}(x,t) = a^2 y_{xx}(x,t) - by_t(x,t) \qquad (b = B\delta^{-1}).$$

If one end $x = 0$ of the string is kept fixed at the origin at all times $t \geq 0$, the boundary condition at that end is clearly

(4)
$$y(0,t) = 0 \qquad\qquad (t \geq 0).$$

But if that end is permitted to slide along the y axis and if the end is moved along that axis with a displacement $f(t)$, the boundary condition is the linear nonhomogeneous condition

$$(5) \qquad\qquad y(0,t) = f(t) \qquad\qquad (t \geqq 0).$$

When the left-hand end is looped around the y axis and a force $g(t)$ in the y direction is applied to that end, $g(t)$ is the limit of the force $V(x,t)$ described in Sec. 3 as x tends to zero through positive values. The boundary condition is then

$$(6) \qquad\qquad -Hy_x(0,t) = g(t) \qquad\qquad (t > 0).$$

The negative sign disappears if $x = 0$ is the right-hand end because $g(t)$ is then the force exerted on the part of the string to the left of that end.

5. Other Examples of Wave Equations. We can present further functions in physics and engineering which satisfy wave

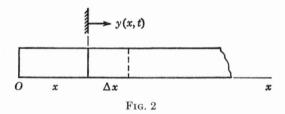

F<small>IG</small>. 2

equations and still limit our attention to fairly simple physical phenomena.

Longitudinal Vibrations of Bars. Let the coordinate x denote the distance from one end of an elastic bar in the shape of a cylinder or prism to other cross sections when the bar is unstrained. Displacements of the ends or initial displacements or velocities of the bar, all directed lengthwise along the bar and uniform over each cross section involved, cause the sections of the bar to move in the direction of the x axis. At time t the longitudinal displacement of the section labeled x is denoted by $y(x,t)$. Thus the origin of the displacement y of that section is fixed outside the bar, in the plane of the original reference position of that section (Fig. 2).

At the same time a neighboring section labeled $x + \Delta x$, to the right of section x, has a displacement $y(x + \Delta x, t)$; thus the element of the bar with natural length Δx is stretched by the amount $y(x + \Delta x, t) - y(x,t)$. Assuming that this extension or compression of the element satisfies Hooke's law, the force exerted upon the element over its left-hand end is, except for the effect of the inertia of the moving element,

$$-AE \frac{y(x + \Delta x,\, t) - y(x,t)}{\Delta x},$$

where A is the area of a cross section and E is the modulus of elasticity of the material in tension and compression. When Δx tends to zero, then it follows that the total longitudinal force $p(x,t)$ exerted on the section x by the part of the bar on the left of that section is given by the basic formula

$$(1) \qquad\qquad p(x,t) = -AEy_x(x,t).$$

Let δ denote the mass of the material per unit volume. When we apply Newton's second law to the motion of an element of the bar of length Δx,

$$(2) \qquad \delta A\, \Delta x\, y_{tt}(x,t) = -AEy_x(x,t) + AEy_x(x + \Delta x,\, t),$$

where the last term represents the force on the element at the end $x + \Delta x$, we find, after dividing by $\delta A\, \Delta x$ and letting Δx tend to zero, that

$$(3) \qquad\qquad y_{tt}(x,t) = a^2 y_{xx}(x,t) \qquad\qquad (a^2 = E\delta^{-1}).$$

Thus the longitudinal displacements $y(x,t)$ in an elastic bar satisfy the wave equation (3) when no external longitudinal forces act on the bar other than at the ends. We have assumed only that displacements are small enough that Hooke's law applies and that sections remain plane after being displaced. The elastic bar here may be replaced by a column of air; then equation (3) has applications in the theory of sound.

The boundary condition $y(0,t) = 0$ signifies that the end $x = 0$ of the bar is held fixed. If instead the end $x = 0$ is free when $t > 0$, then no force acts across that end; that is, $p(0,t) = 0$ or, in view of formula (1),

$$(4) \qquad\qquad y_x(0,t) = 0 \qquad\qquad (t > 0).$$

Transverse Vibrations of Membranes. Let $z(x,y,t)$ denote small displacements in the z direction, at time t, of points $(x,y,0)$ of a flexible membrane stretched tightly over a frame in the xy plane. The tensile stress P, the tension per unit length across any line on the membrane, is large, and the magnitude H of its component parallel to the xy plane is assumed to be constant. Then the internal force in the z direction at a section $x = x_0$, per unit length of that line, is $-Hz_x(x_0,y,t)$, corresponding to the force V (Sec. 3) in the vibrating string. The force in the z direction at a section $y = y_0$, per unit length, is $-Hz_y(x,y_0,t)$.

Consider an element of the membrane whose projection on the xy plane is a rectangle with opposite vertices $(x,y,0)$ and $(x + \Delta x, y + \Delta y, 0)$. When Newton's second law is applied to the motion of that element in the z direction, we find that $z(x,y,t)$ satisfies the two-dimensional wave equation

$$(5) \qquad z_{tt} = a^2(z_{xx} + z_{yy}) \qquad (a^2 = H\delta^{-1}).$$

Here δ is the mass of the membrane per unit area. Details of the derivation are left to the problems.

If an external transverse force $F(x,y,t)$ per unit area acts over the membrane, the equation of motion takes the form

$$(6) \qquad z_{tt} = a^2(z_{xx} + z_{yy}) + F\delta^{-1}.$$

PROBLEMS

1. Give details in the derivation of equation (1), Sec. 4, for the forced vibrations of a stretched string.

2. A tightly stretched string with its ends fixed at the points $(0,0)$ and $(2c,0)$ hangs at rest under its own weight. The y axis points vertically upward. State why the static displacements $y(x)$ of points of the string satisfy the boundary value problem

$$a^2y''(x) - g = 0 \qquad (0 < x < 2c),$$
$$y(0) = y(2c) = 0.$$

Hence show that the string hangs in the parabolic arc

$$(x - c)^2 = \frac{2a^2}{g}\left(y + \frac{gc^2}{2a^2}\right) \qquad (0 \leqq x \leqq 2c).$$

Show that the depth of the vertex of the arc varies directly with δ and c^2 and inversely with H.

3. Use formula (1), Sec. 3, for the vertical force V and the formula for y in Problem 2 to show that the vertical force exerted on that string by either support is $g\delta c$, half the weight of the string.

4. A strand of wire 1 ft long, stretched between the origin and the point $(1,0)$ with tension $H = 10$ lb, weighs 0.032 lb ($g\delta = 0.032$, $g =$

32 ft/sec²). At the instant $t = 0$ the strand lies along the x axis, but it has a velocity 1 ft/sec in the direction of the y axis, perhaps because the supports were in motion and were brought to rest at that instant. If no external forces act along the wire, show why the displacements $y(x,t)$ should satisfy this boundary value problem:

$$y_{tt}(x,t) = 10^4 y_{xx}(x,t) \qquad (0 < x < 1, t > 0),$$
$$y(0,t) = y(1,t) = 0, \qquad y(x,0) = 0, \qquad y_t(x,0) = 1.$$

5. The physical dimensions of the force H, the tension in the string, are those of mass times acceleration, MLT^{-2}, where M denotes mass, L length, and T time. Since $a^2 = H\delta^{-1}$, show that a has the dimensions of velocity LT^{-1}.

6. The end $x = 0$ of a cylindrical elastic bar is kept fixed, and a constant compressive force of magnitude F_0 units per unit area is exerted at all times $t > 0$ over the end $x = c$. If the bar is initially unstrained and at rest and if no external forces act along the bar, verify that the function $y(x,t)$ representing the longitudinal displacements of cross sections should satisfy this boundary value problem:

$$y_{tt}(x,t) = a^2 y_{xx}(x,t) \qquad (0 < x < c, t > 0; a^2 = E\delta^{-1})$$
$$y(0,t) = 0, \qquad E y_x(c,t) = -F_0, \qquad y(x,0) = y_t(x,0) = 0.$$

7. The left-hand end $x = 0$ of an elastic bar is elastically supported in such a way that the longitudinal force per unit area exerted on the bar at that end is proportional to the displacement of the end, but opposite in sign. Show that the end condition there has the form

$$E y_x(0,t) = K y(0,t) \qquad\qquad (K > 0).$$

8. Derive equation (6), Sec. 5. Also note that the *static transverse displacements* $z(x,y)$ of a membrane, over which a transverse force $F(x,y)$ per unit area acts, satisfy *Poisson's equation*

$$z_{xx} + z_{yy} + k = 0 \qquad\qquad (k = FH^{-1}).$$

6. Conduction of Heat.
Thermal energy is transferred from warmer to cooler regions interior to a solid body by conduction. It is convenient to refer to that transfer as the flow of heat, as if heat were a fluid or gas which diffuses through the body from regions of high concentration into regions of low concentration of that fluid.

Let P_0 denote a point (x_0,y_0,z_0) interior to the body and S a plane or smooth curved surface through P_0. At a time t_0 the *flux* $\Phi(x_0,y_0,z_0,t_0)$ *of heat* across S at P_0 is the quantity of heat per unit area, per unit time, that is being conducted across S at that point. Flux is measured in such units as calories per square centimeter per second.

If $u(x,y,z,t)$ denotes the temperatures at points of the solid at time t and if n is a coordinate that represents distance normal to surface S at the interior point P_0 (Fig. 3), the flux across S will be in the positive direction of n if du/dn is negative. A *fundamental postulate of the mathematical theory of heat conduction* states that there is a thermal coefficient K of the material such that

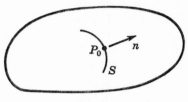

Fig. 3

$$(1) \qquad\qquad \Phi = -K\frac{du}{dn} \qquad\qquad (K > 0);$$

that is, the flux $\Phi(x_0,y_0,z_0,t_0)$ across S is proportional to the value of the directional derivative of the temperature function u in the direction normal to S, at P_0, t_0. The coefficient K is called the *thermal conductivity* of the material.

In particular, the flux across each of the planes $x = x_0$, $y = y_0$, and $z = z_0$ through the point P_0 has the values

$$(2) \qquad \Phi_1 = -Ku_x, \qquad \Phi_2 = -Ku_y, \qquad \Phi_3 = -Ku_z,$$

respectively, when the partial derivatives are evaluated at P_0, t_0. For a fixed value of t, u_x, u_y, and u_z are the components of the gradient of u. Since the direction of the vector grad u is that in which u increases most rapidly, the vector

$$(3) \qquad\qquad \mathbf{J} = -K \text{ grad } u$$

is called the total flux or the current of heat flow at the point. It measures the flux across the isothermal surface

$$u(x,y,z,t_0) = u(x_0,y_0,z_0,t_0)$$

through P_0. The projection $J_{(n)}$ of $\mathbf{J}$ on the normal to the arbitrary surface S is the flux Φ given by formula (1):

$$J_{(n)} = \Phi = -K \, du/dn$$

Another thermal coefficient of the material is the *heat capacity c_0 per unit volume.* This is the quantity of heat required to raise the temperature of a unit volume of the material by a unit on the temperature scale.

Unless otherwise stated, *we assume here that the coefficients c_0 and K are constants.* In this case a *second postulate* of the mathematical theory is that conduction leads to a temperature function u which, together with its derivative u_t and its derivatives of first and second order with respect to x, y, and z, are all continuous functions throughout each domain interior to the solid in which no heat is generated.

To derive the differential equation satisfied by u, first consider a spherical surface S_0 with center at P_0, small enough that all points within and on S_0 are interior to the solid body. The integral of $c_0 u(x,y,z,t)$ over the region R_0 bounded by S_0 is a measure of the instantaneous heat content $Q_0(t)$ of R_0. Hence

$$(4) \qquad Q_0'(t) = \frac{d}{dt} \iiint_{R_0} c_0 u \, dV = c_0 \iiint_{R_0} \frac{\partial u}{\partial t} \, dV,$$

where dV is written for the volume element in the integral.

Heat enters R_0 only by conduction through its boundary surface S_0. If n is distance normal to that surface, positive in the outward direction, and if dA denotes the element of area of S_0, then the time rate $Q_0'(t)$ of conduction of heat *into* the sphere has the alternate expression

$$(5) \qquad Q_0'(t) = \iint_{S_0} K \frac{du}{dn} \, dA = - \iint_{S_0} J_{(n)} \, dA.$$

According to the divergence theorem for transforming surface integrals into volume integrals,

$$(6) \qquad \iint_{S_0} J_{(n)} \, dA = \iiint_{R_0} \operatorname{div} \mathbf{J} \, dV.$$

Thus formula (5) can be written

$$(7) \quad Q_0'(t) = - \iiint_{R_0} \operatorname{div} \mathbf{J} \, dV = K \iiint_{R_0} \operatorname{div} (\operatorname{grad} u) \, dV.$$

Now div (grad u) is the laplacian

$$\nabla^2 u = u_{xx} + u_{yy} + u_{zz};$$

so it follows from expressions (4) and (7) that

$$(8) \qquad \iiint_{R_0} (c_0 u_t - K \nabla^2 u) \, dV = 0.$$

Equation (8) holds for each spherical region R_0 about P_0, provided only that all points of R_0 are interior to the solid and that R_0 is free from sources. The integrand is a continuous function of x, y, and z in R_0, at time t. If the integrand has a positive value at P_0, its continuity requires that there be a spherical region R_1 about P_0, where R_1 is interior to R_0, such that $c_0 u_t - K\nabla^2 u > 0$ throughout R_1. The integral in equation (8), with R_0 replaced by R_1, would then have a positive value rather than the value zero. The corresponding contradiction arises if we assume that the integrand has a negative value at P_0. Consequently at P_0

$$c_0 u_t - K\nabla^2 u = 0.$$

Since P_0 represents any interior point of the body in a domain free from sources, the temperature function $u(x,y,z,t)$ in every such domain satisfies the *heat equation*

$$(9) \qquad u_t = k(u_{xx} + u_{yy} + u_{zz}),$$

where k is the *thermal diffusivity* of the material

$$(10) \qquad k = \frac{K}{c_0}.$$

7. Discussion of the Heat Equation. If the temperatures within a solid are independent of z, that is, if there is no flow of heat in the z direction, the heat equation reduces to the equation for two-dimensional flow parallel to the xy plane:

$$(1) \qquad u_t = k(u_{xx} + u_{yy}).$$

For one-dimensional flow parallel to the x axis, the equation becomes

$$(2) \qquad u_t(x,t) = k u_{xx}(x,t).$$

When temperatures are in a *steady state*, that is, when u does not vary with time, the heat equation becomes *Laplace's equation* $\nabla^2 u = 0$:

$$(3) \qquad u_{xx} + u_{yy} + u_{zz} = 0.$$

We have assumed so far that heat is neither generated nor lost within the solid but only transferred by conduction. In case there is a uniform source throughout the solid that generates heat at a constant rate q, where q denotes quantity of heat generated per unit volume per unit time, the heat equation takes the form

$$(4) \qquad c_0 u_t = K\nabla^2 u + q.$$

The derivation of the heat equation in Sec. 6 can be modified to obtain formula (4) if the postulate on the continuity of u and its derivatives is extended to include cases in which a constant uniform source is present.

Boundary conditions which describe the thermal conditions on the surface of the solid and the initial temperature distribution throughout the solid must accompany the heat equation if we are to determine the temperature function u. If the plane $x = 0$ is a boundary surface that is perfectly insulated, for example, the boundary condition there is

$$(5) \qquad u_x(0,y,z,t) = 0$$

because $-Ku_x(0,y,z,t)$ is the flux of heat across that plane.

The postulate $\Phi = -K\,du/dn$ also applies to simple diffusion to represent the flux of a substance which is diffusing within a porous solid. In this case the function u denotes *concentration* (the mass of the diffusing substance per unit volume of the solid), and K is the *coefficient of diffusion*. Since the content of the substance in a given region is the integral of u over that region, we can replace c_0 by unity in the derivation in Sec. 6 to see that the concentration u satisfies the *equation of diffusion*

$$(6) \qquad u_t = K\nabla^2 u.$$

PROBLEMS

1. Let $u(x)$ denote the steady-state temperatures in a slab bounded by the planes $x = 0$ and $x = b$, when those boundaries are kept at fixed temperatures $u = 0$ and $u = u_0$, respectively. Set up the boundary value problem for $u(x)$ and solve it to show that

$$u(x) = \frac{u_0}{b}\,x, \qquad \Phi = -K\frac{u_0}{b},$$

where Φ is the flux of heat across each plane $x = x_0$ $(0 \le x_0 \le b)$.

2. A slab occupies the region $0 \le x \le b$. There is a constant flux of heat $\Phi = \phi_0$ into the slab through the face $x = 0$. The face $x = b$ is kept at temperature $u = 0$. Set up and solve the boundary value problem for the steady-state temperature $u(x)$ in the slab.

$$Ans.\ u(x) = \frac{\phi_0}{K} (b - x).$$

3. Let the slab $0 \leq x \leq b$ be subjected to *surface heat transfer* to surrounding media at its faces $x = 0$ and $x = b$, according to the following linear law. The flux out of each face is proportional to the difference between the temperature of that face and the temperature of the adjoining medium. Let the positive constant H be the proportionality factor for both faces. If the medium $x < 0$ has temperature zero and the medium $x > b$ has the constant temperature T_1, verify that the boundary value problem for steady-state temperatures $u(x)$ in the slab is

$$u''(x) = 0 \qquad (0 < x < b),$$
$$Ku'(0) = Hu(0), \qquad -Ku'(b) = H[u(b) - T_1].$$

Write $h = H/K$ and show that

$$u(x) = \frac{T_1}{bh + 2} (hx + 1).$$

4. When heat is generated at a rate q per unit volume per unit time throughout a solid, point out why equation (5) of Sec. 6 should be modified to read

$$Q_0'(t) = - \iint_{S_0} J_{(n)}\ dA + \iiint_{R_0} q\ dV.$$

Let q be a constant or a continuous function of x, y, z. Then we postulate the continuity of u and its derivatives as before. Complete the derivation of the form (4), Sec. 7, of the heat equation when q is such a function.

5. Let $u(x,t)$ denote temperatures within a solid which is free from sources. Consider a cylindrical element interior to the solid such that the axis of the cylinder is parallel to the x axis and of length Δx. If the area of the base of the cylinder is A, show why

$$A[-Ku_x(x,t) + Ku_x(x + \Delta x, t)] = c_0 A\ \Delta x\ u_t(x,t)$$

approximately if Δx is small, and thus obtain equation (2), Sec. 7.

6. In Problem 5, let heat be generated at a constant rate q per unit volume per unit time throughout the solid and derive the equation

$$c_0 u_t = Ku_{xx} + q.$$

7. Let $u(x,t)$ denote temperatures in a slender wire lying along the x axis when heat transfer takes place along the wire into the surrounding medium at fixed temperature T_0. If the time rate of transfer per unit length is proportional to $u(x,t) - T_0$, use the procedure indicated in Problem 5 to derive the equation

$$u_t = ku_{xx} - h(u - T_0) \qquad (h \text{ constant}, h > 0).$$

8. When the thermal coefficients c_0 and K are functions of x, y, z, and t, modify the derivation in Sec. 6 to show that the heat equation takes the form

$$(c_0 u)_t = (Ku_x)_x + (Ku_y)_y + (Ku_z)_z$$

in a domain where all functions and derivatives involved in that equation are continuous.

8. Laplace's Equation.　A function $u(x,y,z)$ that is continuous, together with its partial derivatives of the first and second order, and satisfies Laplace's equation

(1) $$\nabla^2 u = 0,$$

where ∇^2 is the laplacian operator

(2) $$\nabla^2 u = \text{div} (\text{grad } u) = u_{xx} + u_{yy} + u_{zz},$$

is called a *harmonic function*.

We have seen that the steady-state temperatures at points interior to a solid in which no heat is generated are represented by a harmonic function u. In fact the temperature function u serves as a potential function for the flow of heat, in the sense that $-K$ grad u represents the current of heat flow by conduction, and thus $-K \, du/dn$ represents the flux of heat in the direction of the coordinate n.

The steady-state concentration of a diffusing substance (Sec. 7) is likewise represented by a harmonic function. From equation (5), Sec. 5, we can see that the static transverse displacements $z(x,y)$ of a stretched membrane satisfy Laplace's equation in two dimensions. Here the displacements are the result of displacements perpendicular to the xy plane of parts of the frame that supports the membrane when no external forces are exerted except at the boundary.

Among the many physical examples of harmonic functions the velocity potential for the steady-state irrotational motion of an incompressible fluid is a prominent one in hydrodynamics and aerodynamics.[1]

---

[1] See chap. 9 of the author's book "Complex Variables and Applications," listed in the Bibliography at the end of our book.

An important harmonic function in electrical field theory is the electrostatic potential $V(x,y,z)$ in a region of space that is free from electric charges. The potential may be produced by any static distribution of electric charges outside that region. The vector $-$ grad V represents the electrostatic intensity at (x,y,z), the electrical force that would be exerted upon a unit positive charge placed at that point. The fact that V is harmonic is a consequence of the inverse-square law of attraction or repulsion between electric charges.

Likewise, gravitational potential is a harmonic function in domains of space not occupied by matter.

The physical problems treated in this book can be limited to those for which the differential equations were derived in this chapter. Equations satisfied by a few other functions in the physical sciences are noted in order to indicate the great variety of applications of partial differential equations. Derivations of equations for other functions will be found in books on hydrodynamics, elasticity, vibrations and sound, electrical field theory, theory of potential, and other branches of mechanics of the continuum.

9. Cylindrical and Spherical Coordinates. The cylindrical coordinates (ρ,ϕ,z), shown in Fig. 4, determine a point P whose rectangular cartesian coordinates are

$$(1) \qquad x = \rho \cos \phi, \qquad y = \rho \sin \phi, \qquad z = z.$$

Thus ρ and ϕ are the polar coordinates in the xy plane of the point Q, where Q is the projection of P on that plane.

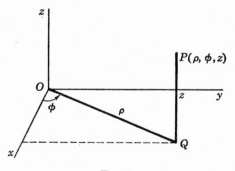

Fig. 4

The relations (1) can be written

$$(2) \qquad \rho = \sqrt{x^2 + y^2}, \qquad \phi = \arctan\frac{y}{x}, \qquad z = z,$$

where the quadrant to which the angle ϕ belongs is determined by the signs of x and y, not by the ratio y/x alone.

Let u denote a function of x, y, and z. Then in view of equations (1) it is also a function of the three independent variables ρ, ϕ, and z. If u and its derivatives with respect to those variables, of first and second order, are continuous functions, we can obtain the laplacian of u in terms of those variables by using the chain rule for differentiating composite functions.

In view of equations (2),

$$(3) \qquad \frac{\partial u}{\partial x} = \frac{\partial u}{\partial \rho}\frac{\partial \rho}{\partial x} + \frac{\partial u}{\partial \phi}\frac{\partial \phi}{\partial x} = \frac{\partial u}{\partial \rho}\frac{x}{\sqrt{x^2 + y^2}} - \frac{\partial u}{\partial \phi}\frac{y}{x^2 + y^2},$$

where y and z are kept fixed in the differentiation $\partial/\partial x$, while ϕ and z are kept fixed in the differentiation $\partial/\partial \rho$, and so on, and ϕ is measured in radians. Similarly,

$$(4) \qquad \frac{\partial u}{\partial y} = \frac{\partial u}{\partial \rho}\frac{y}{\sqrt{x^2 + y^2}} + \frac{\partial u}{\partial \phi}\frac{x}{x^2 + y^2}.$$

From formula (3) we can write

$$\frac{\partial^2 u}{\partial x^2} = \frac{\partial u}{\partial \rho}\frac{\partial}{\partial x}\left(\frac{x}{\rho}\right) - \frac{\partial u}{\partial \phi}\frac{\partial}{\partial x}\left(\frac{y}{\rho^2}\right) + \frac{x}{\rho}\frac{\partial}{\partial x}\left(\frac{\partial u}{\partial \rho}\right) - \frac{y}{\rho^2}\frac{\partial}{\partial x}\left(\frac{\partial u}{\partial \phi}\right).$$

The chain rule applies to the last two indicated derivatives, giving

$$\frac{\partial}{\partial x}\left(\frac{\partial u}{\partial \rho}\right) = \frac{\partial^2 u}{\partial \rho^2}\frac{x}{\rho} - \frac{\partial^2 u}{\partial \phi\,\partial \rho}\frac{y}{\rho^2},$$

$$\frac{\partial}{\partial x}\left(\frac{\partial u}{\partial \phi}\right) = \frac{\partial^2 u}{\partial \rho\,\partial \phi}\frac{x}{\rho} - \frac{\partial^2 u}{\partial^2 \phi}\frac{y}{\rho^2}.$$

Substituting and simplifying, we find that

$$(5) \qquad \frac{\partial^2 u}{\partial x^2} = \frac{y^2}{\rho^3}\frac{\partial u}{\partial \rho} + \frac{2xy}{\rho^4}\frac{\partial u}{\partial \phi} + \frac{x^2}{\rho^2}\frac{\partial^2 u}{\partial \rho^2} - \frac{2xy}{\rho^3}\frac{\partial^2 u}{\partial \rho\,\partial \phi} + \frac{y^2}{\rho^4}\frac{\partial^2 u}{\partial \phi^2}.$$

In like manner we find from equation (4) that

$$(6) \qquad \frac{\partial^2 u}{\partial y^2} = \frac{x^2}{\rho^3}\frac{\partial u}{\partial \rho} - \frac{2xy}{\rho^4}\frac{\partial u}{\partial \phi} + \frac{y^2}{\rho^2}\frac{\partial^2 u}{\partial \rho^2} + \frac{2xy}{\rho^3}\frac{\partial^2 u}{\partial \rho\,\partial \phi} + \frac{x^2}{\rho^4}\frac{\partial^2 u}{\partial \phi^2}.$$

Hence the *laplacian of u in cylindrical coordinates is*

(7) $$\nabla^2 u = \frac{\partial^2 u}{\partial \rho^2} + \frac{1}{\rho}\frac{\partial u}{\partial \rho} + \frac{1}{\rho^2}\frac{\partial^2 u}{\partial \phi^2} + \frac{\partial^2 u}{\partial z^2}.$$

We may group the first two terms and use the subscript notation for partial derivatives to write this in the form

(8) $$\nabla^2 u = \rho^{-1}(\rho u_\rho)_\rho + \rho^{-2}u_{\phi\phi} + u_{zz}.$$

The *spherical coordinates* (r,ϕ,θ) of a point (Fig. 5) are related to x, y, and z as follows:

(9) $x = r\sin\theta\cos\phi$, $y = r\sin\theta\sin\phi$, $z = r\cos\theta$.

The coordinate ϕ is common to cylindrical and spherical coordinates, while the other coordinates in the two systems are

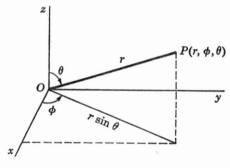

Fig. 5

related by equations of the same form as equations (1), namely,

(10) $$z = r\cos\theta,\qquad \rho = r\sin\theta.$$

Consequently the expression (7) can be transformed to spherical coordinates by steps that correspond to those used above or by the proper interchange of letters.   It turns out that the *laplacian in spherical coordinates* can be written

(11) $$\nabla^2 u = \frac{\partial^2 u}{\partial r^2} + \frac{2}{r}\frac{\partial u}{\partial r} + \frac{1}{r^2\sin^2\theta}\frac{\partial^2 u}{\partial \phi^2} + \frac{1}{r^2}\frac{\partial^2 u}{\partial \theta^2} + \frac{\cot\theta}{r^2}\frac{\partial u}{\partial \theta}.$$

Another form of this expression is

(12) $$\nabla^2 u = \frac{1}{r}(ru)_{rr} + \frac{1}{r^2\sin^2\theta}u_{\phi\phi} + \frac{1}{r^2\sin\theta}(u_\theta\sin\theta)_\theta.$$

PROBLEMS

1. Derive formulas (4) and (6) of this section.

2. Given that $u(x,y,z)$ and its partial derivatives of first and second order are continuous functions, transform the right-hand member of formula (7) into $u_{xx} + u_{yy} + u_{zz}$.

3. Show that formula (12) is the same as (11).

4. Derive formula (11) from formula (7). *Suggestions:* When ϕ is fixed, u is a function of the independent variables z and ρ or, in view of equations (10), of r and θ. First obtain the formula

$$u_\rho = u_r \rho r^{-1} + u_\theta z r^{-2}$$

corresponding to equation (4), then the formula

$$u_{zz} + u_{\rho\rho} = u_{rr} + r^{-1}u_r + r^{-2}u_{\theta\theta},$$

which corresponds to formula (7) when $u = u(x,y)$.

5. Let $u(r)$ denote the steady-state temperatures in a solid bounded by two concentric spheres $r = 1$ and $r = r_0$ $(0 \leq \phi < 2\pi, 0 \leq \theta \leq \pi)$, when $u = 0$ on the inner surface $r = 1$ and $u = u_0$ on the outer surface $r = r_0$, where u_0 is a constant. Note why the heat equation for $u(r)$ reduces to

$$\frac{d^2}{dr^2}(ru) = 0$$

and obtain the formula

$$u(r) = \frac{r_0 u_0}{r_0 - 1}\left(1 - \frac{1}{r}\right) \qquad (1 \leq r \leq r_0).$$

6. Let $z(\rho)$ represent the static transverse displacements of a membrane stretched between two circles $\rho = 1$ and $\rho = \rho_0$ in the plane $z = 0$ after the outer support $\rho = \rho_0$ is displaced by a distance $z = z_0$. State why the boundary value problem in $z(\rho)$ can be written

$$\frac{d}{d\rho}\left(\rho \frac{dz}{d\rho}\right) = 0 \qquad (1 < \rho < \rho_0),$$
$$z(1) = 0, \qquad z(\rho_0) = z_0,$$

and obtain the formula

$$z(\rho) = z_0 \frac{\log \rho}{\log \rho_0} \qquad (1 \leq \rho \leq \rho_0).$$

7. State why the steady-state temperatures $u(\rho)$ in a hollow cylinder $1 \leq \rho \leq \rho_0$, $-\infty < z < \infty$ also satisfies the boundary value problem written in Problem 6 if $u = 0$ on the inner cylindrical surface and $u = z_0$ on the surface $\rho = \rho_0$. Thus Problem 6 is a *membrane analogy* for this problem. Soap films have been used to display such analogies.

8. A uniform transverse force, F_0 units per unit area, acts over a membrane stretched between the circles $\rho = 1$, $z = 0$ and $\rho = \rho_0$, $z = 0$. From Problem 8, Sec. 5, show that the static transverse displacements $z(\rho)$ satisfy the equation

$$(\rho z')' + k_0\rho = 0 \qquad\qquad (k_0 = F_0H^{-1}),$$

and derive the formula

$$z(\rho) = \frac{k_0}{4}\,(\rho_0{}^2 - 1)\left(\frac{\log \rho}{\log \rho_0} - \frac{\rho^2 - 1}{\rho_0{}^2 - 1}\right) \qquad (1 \leqq \rho \leqq \rho_0).$$

10. Types of Equations and Conditions. The linear partial differential equation of second order in $u(x,y)$,

$$(1) \qquad A u_{xx} + B u_{xy} + C u_{yy} + D u_x + E u_y + F u = G,$$

where A, B, . . . , G are constants or functions of x and y only, is of *elliptic, parabolic, or hyperbolic type* in a domain of the xy plane if the values of the function $B^2 - 4AC$ are negative, zero, or positive, respectively, throughout the domain. The three types require different kinds of boundary conditions to determine a solution.

Poisson's equation in $u(x,y)$, $\nabla^2 u = G$, and Laplace's equation are special cases of equation (1) in which $A = C = 1$ and B, D, E, and F vanish. Hence those equations are elliptic in every domain. The heat equation $k u_{xx} - u_y = 0$ is parabolic, and the wave equation $u_{xx} - a^2 u_{yy} = 0$ is hyperbolic.

In the theory of partial differential equations[1] it is shown that equation (1) can be reduced to generalized forms of Poisson's equation or the heat equation or the wave equation, according as equation (1) is elliptic, parabolic, or hyperbolic. The reductions are made by substituting new independent variables for x and y.

Another special case of equation (1) is the *telegraph equation*

$$(2) \qquad v_{xx} = KLv_{tt} + (KR + LS)v_t + RSv.$$

Here $v(x,t)$ represents either the electric potential or current at time t at a point x units from one end of a transmission line or cable which has electrostatic capacity K, self-inductance L, resistance R, and leakage conductance S, all per unit length. The

---

[1] See, for instance, books on partial differential equations by Courant and Hilbert, vol. 2, Frank and v. Mises, or Greenspan listed in the Bibliography.

equation is hyperbolic if $KL > 0$.   It is parabolic in case either K or L vanishes.

Let u denote the unknown function in a boundary value problem.   A condition that prescribes the values of u itself along a boundary is known as a boundary condition of *Dirichlet type*. A problem of determining a harmonic function interior to a region so that the function assumes prescribed values over the boundary of the region is a *Dirichlet problem*.   In this case the values of the function can be interpreted as steady-state temperatures and this physical interpretation leads us to expect that a Dirichlet problem may have a unique solution if the functions considered satisfy certain requirements as to their regularity.

A boundary condition of the *second type*, called a *Neumann condition*, prescribes the values of the normal derivative du/dn of the function at the boundary.   Among other kinds of boundary conditions are those of *third type* in which values of $hu + du/dn$ are prescribed at the boundary, where h is either a constant or a function of the independent variables.

If the partial differential equation in u is of second order with respect to one of the independent variables t and if the values of both u and u_t are prescribed on a boundary $t = 0$, the boundary condition is one of *Cauchy type* with respect to t.   When the differential equation is the wave equation $u_{tt} = a^2 u_{xx}$, such a condition corresponds physically to that of prescribing the initial values of both the transverse displacements u and velocities u_t of a stretched string.   Both conditions appear to be needed if the displacements $u(x,t)$ are to be determined.

When the equation is Laplace's equation $u_{xx} = -u_{yy}$ or the heat equation $ku_{xx} = u_t$, however, conditions of Cauchy type with respect to x cannot be imposed without severe restrictions. This again is suggested by interpreting u physically, as a temperature function.   When the temperature u is prescribed on the boundary $x = 0$, the flux Ku_x through that boundary is ordinarily determined by the values of u there and by other conditions in the problem.   Conversely, if the flux $Ku_x(0,t)$ is prescribed, specific temperatures $u(0,t)$ are needed to produce that flux.

CHAPTER 2

SUPERPOSITION OF SOLUTIONS

11. Linear Combinations. If c_1 and c_2 are constants and u_1 and u_2 are functions, then the function

$$c_1 u_1 + c_2 u_2$$

is called a *linear combination* of the functions u_1 and u_2. Note that $c_1 u_1$, $c_2 u_2$, and $u_1 \pm u_2$ are special cases.

According to elementary properties of derivatives, a derivative of each linear combination of two functions can be written as the linear combination of derivatives of the functions:

$$(1) \qquad \frac{\partial}{\partial x}(c_1 u_1 + c_2 u_2) = c_1 \frac{\partial u_1}{\partial x} + c_2 \frac{\partial u_2}{\partial x}.$$

A *linear operator* on functions is an operator L that transforms each function u of some class of functions into a function Lu and has this property: for each pair of functions u_1 and u_2 of that class it is true that

$$(2) \qquad L(c_1 u_1 + c_2 u_2) = c_1 L u_1 + c_2 L u_2,$$

whenever c_1 and c_2 are constants. In particular then,

$$L(c_1 u_1) = c_1 L u_1, \qquad L(u_1 + u_2) = L u_1 + L u_2, \qquad L(0) = 0.$$

Property (2) can be extended immediately; for if u_3 is a third function of that class, then

$$L(c_1 u_1 + c_2 u_2 + c_3 u_3) = L(c_1 u_1 + c_2 u_2) + c_3 L u_3$$
$$= c_1 L u_1 + c_2 L u_2 + c_3 L u_3.$$

Proceeding by induction we find that L transforms linear combinations of n functions in this manner:

$$(3) \qquad L\left(\sum_{i=1}^{n} c_i u_i\right) = \sum_{i=1}^{n} c_i L u_i.$$

23

In view of equation (1), $\partial/\partial x$ is a linear operator on functions that have derivatives of the first order with respect to x. It is naturally classified as a linear *differential* operator. The operator $f(x,y)\times$ which transforms each function $u(x,y)$ into the product fu is an example of a still more elementary linear operator. Let us now point out why such composites as $f\partial/\partial x$, $\partial^2/\partial x^2$, and $f\partial/\partial x + \partial^2/\partial x^2$ are also linear operators.

Given two linear operators L and M, distinct or not, such that M transforms each function u of some class into a function Mu to which L applies, then if u_1 and u_2 are functions of that class, it follows from equation (2) that

$$(4)\quad LM(c_1u_1 + c_2u_2) = L(c_1Mu_1 + c_2Mu_2) = c_1LMu_1 + c_2LMu_2.$$

In brief, the *product* LM of linear operators is a linear operator. In case L and M both represent $\partial/\partial x$, it follows that $\partial^2/\partial x^2$ is a linear operator for the class of all functions u whose derivatives u_x and u_{xx} exist. Similarly if M is $\partial/\partial x$ and L is $f(x,y)\times$, it follows that $f\partial/\partial x$ is a linear operator on $u(x,y)$.

The *sum* of two linear operators is defined by the equation

$$(5)\qquad\qquad (L + M)u = Lu + Mu.$$

If we replace u here by $c_1u_1 + c_2u_2$, we can see that the sum $L + M$ is a linear operator and hence that the sum of any finite number of linear operators is linear.

Each term of a linear homogeneous differential equation in u consists of a product of a function of the independent variables by one of the derivatives of u or by u itself. Hence every linear homogeneous differential equation has the form

$$(6)\qquad\qquad\qquad Lu = 0,$$

where L is a linear differential operator. In case

$$(7)\quad L = A\,\frac{\partial^2}{\partial x^2} + B\,\frac{\partial^2}{\partial x\,\partial y} + C\,\frac{\partial^2}{\partial y^2} + D\,\frac{\partial}{\partial x} + E\,\frac{\partial}{\partial y} + F\times,$$

for example, where the letters A to F denote functions of x and y only, equation (6) is the linear homogeneous partial differential equation in $u(x,y)$ of the second order.

Linear homogeneous boundary conditions also have the form (6). Then the variables appearing as arguments of u and as arguments of functions that serve as coefficients in the linear operator L are restricted so that they represent points on the boundary.

Now let u_i denote functions that satisfy equation (6); that is, $Lu_i = 0$ $(i = 1, 2, \ldots, n)$. Then it follows from equation (3) that each linear combination of those functions also satisfies that equation. We state that *principle of superposition* of solutions, fundamental to one of the most powerful methods of solving linear boundary value problems, as follows.

Theorem 1. *If each of n functions u_1, u_2, $\ldots$, u_n satisfies a linear homogeneous differential equation, then every linear combination of those functions,*

$$(8) \qquad u = c_1u_1 + c_2u_2 + \cdots + c_nu_n,$$

where the c's are arbitrary constants, satisfies that differential equation. If each of the n functions satisfies a linear homogeneous boundary condition, then every linear combination (8) satisfies that boundary condition.

12. Examples. The principle of superposition is useful in ordinary differential equations. For example, from the particular solutions $y = e^x$ and $y = e^{-x}$ of the linear homogeneous equation $y'' - y = 0$, the general solution $y = c_1e^x + c_2e^{-x}$ can be written.

To illustrate the application of Theorem 1 to partial differential equations, consider the linear homogeneous heat equation

$$(1) \qquad u_t(x,t) = u_{xx}(x,t)$$

and the two linear homogeneous boundary conditions

$$(2) \qquad u_x(0,t) = 0, \qquad u_x(1,t) = 0.$$

It is easy to verify that each of the functions

$$u_0 = 1, \qquad u_n = \exp(-n^2\pi^2 t) \cos n\pi x \quad (n = 1, 2, \ldots),$$

where $\exp r$ denotes the exponential function e^r, satisfies equation (1) and conditions (2). Hence every linear combination

$$(3) \qquad u = c_0 + \sum_{n=1}^{m} c_n \exp(-n^2\pi^2 t) \cos n\pi x$$

of those functions satisfies equation (1) and conditions (2).

A principle of superposition for nonhomogeneous linear differential equations should be noted. The function

$$(4) \qquad\qquad u = u_1 + u_2$$

clearly satisfies the linear differential equation

$$(5) \qquad\qquad Lu = f_1 + f_2,$$

where f_1 and f_2 are functions of the independent variables only, if

$$(6) \qquad\quad Lu_1 = f_1 \quad \text{and} \quad Lu_2 = f_2,$$

where L is the same linear differential operator as in equation (5). Note that in case f_1 is identically zero, then u_1 is a solution of the homogeneous equation $Lu_1 = 0$.

The principle is also valid when equations (5) and (6) represent linear boundary conditions.

13. Series of Solutions. In order to extend our results on linear combinations of solutions to an infinite set of functions $u_1, u_2, \ldots$, we must deal with convergence and differentiability of infinite series of functions.

Let the constants c_n and the functions u_n be such that the infinite series with terms $c_n u_n$ converges throughout some domain of the independent variables. The sum of that series is a function u,

$$(1) \qquad\qquad u = \sum_{n=1}^{\infty} c_n u_n.$$

If x represents one of the independent variables, the series is *differentiable*, or *termwise differentiable*, with respect to x if its terms are such that the derivatives $\partial u_n/\partial x$ and $\partial u/\partial x$ exist and the series of the functions $c_n \partial u_n/\partial x$ converges to $\partial u/\partial x$; that is, if

$$(2) \qquad\qquad \frac{\partial u}{\partial x} = \sum_{n=1}^{\infty} c_n \frac{\partial u_n}{\partial x}.$$

Note that a series must necessarily be convergent to be differentiable. Sufficient conditions for differentiability will be noted in Sec. 14.

If, in addition, series (2) is differentiable with respect to x, then series (1) is differentiable twice with respect to x, and so on, for other derivatives.

Let f denote a function of the independent variables. According to the definition of the sum of an infinite series, it follows from equation (2) that

$$(3) \qquad f\frac{\partial u}{\partial x} = f \lim_{m\to\infty} \sum_{n=1}^{m} c_n \frac{\partial u_n}{\partial x} = \lim_{m\to\infty} f \frac{\partial}{\partial x} \sum_{n=1}^{m} c_n u_n.$$

Here the operator $\partial/\partial x$ can be replaced by other derivatives if the series is so differentiable. Let L be a linear operator consisting of a function f times a derivative, or a sum of a finite number of such terms. Then by adding corresponding members of equations similar to equation (3) we find that

$$(4) \qquad Lu = \lim_{m\to\infty} L\left(\sum_{n=1}^{m} c_n u_n\right).$$

The sum in the right-hand member of equation (4) is a linear combination of the functions $u_1, u_2, \ldots, u_m$. If each of the functions u_n satisfies the linear homogeneous differential equation $Lu_n = 0$, it follows from Theorem 1 that

$$L\left(\sum_{n=1}^{m} c_n u_n\right) = 0$$

for every m; hence $Lu = 0$, in view of equation (4).

A linear homogeneous boundary condition may also be represented by an equation $Lu = 0$. In this case we may require the function Lu to satisfy a condition of continuity at points of the boundary in order that its values there will represent limiting values as those points are approached from the interior of the domain.

The following generalization of Theorem 1 is now established.

Theorem 2. *If each function of an infinite set $u_1, u_2, \ldots$ satisfies a linear homogeneous differential equation or boundary condition $Lu = 0$, then the function*

$$(5) \qquad u = \sum_{n=1}^{\infty} c_n u_n$$

also satisfies $Lu = 0$, provided the constants c_n and the functions u_n are such that the series converges and is differentiable, for all derivatives involved in L, and such that required continuity conditions at the boundary are satisfied by Lu when $Lu = 0$ is a boundary condition.

14. Uniform Convergence. Example.   We recall some facts about uniformly convergent series of functions.[1]

Let $S(x)$ denote the sum of an infinite series of functions $v_n(x)$, convergent for all x on some interval,

$$(1) \qquad S(x) = \sum_{n=1}^{\infty} v_n(x) = \lim_{m \to \infty} S_m(x) \qquad (a \leqq x \leqq b),$$

where $S_m(x)$ is the partial sum consisting of the first m terms of the series.   The series converges *uniformly* with respect to x if the absolute value of its remainder $R_m(x) = S(x) - S_m(x)$ can be made arbitrarily small for all x on the interval, by taking m sufficiently large; that is, if for each small positive number ϵ, there exists an integer m_ϵ *independent of x* such that

$$(2) \qquad |S(x) - S_m(x)| < \epsilon \qquad \text{whenever } m > m_\epsilon \quad (a \leqq x \leqq b).$$

A useful *sufficient condition* for uniform convergence is given by the *Weierstrass M-test:* if there is a convergent series of positive constants, $\sum_{n=1}^{\infty} M_n$, such that

$$(3) \qquad |v_n(x)| \leqq M_n \qquad (a \leqq x \leqq b),$$

then series (1) is uniformly convergent on that interval.

Because $S(x) = S_m(x) + R_m(x)$ and $|R_m(x)|$ is uniformly small when m is sufficiently large, the following properties of uniformly convergent series can be established.

If the functions $v_n(x)$ are continuous and if series (1) is uniformly convergent, then the sum $S(x)$ of that series is also a continuous function; and that series can be integrated term by term over the interval to give the integral of $S(x)$ over the interval.

If the functions v_n and their derivatives v_n' are continuous, if

---

[1] See, for instance, W. Kaplan, "Advanced Calculus," pp. 338ff., 1952.

series (1) converges, and if the series whose terms are $v_n'(x)$ is uniformly convergent, then series (1) is differentiable with respect to x.

Corresponding results hold for series of functions of two or more independent variables.

Example. Let us use the above results and Theorem 2 to write a more general solution for the example on the heat equation introduced in Sec. 12. To be specific, we now require the function u to satisfy the equation

$$(4) \qquad u_{xx}(x,t) - u_t(x,t) = 0$$

in the domain $0 < x < 1$, $t > t_0$ whenever $t_0 > 0$ and these two conditions of Neumann type at the boundaries $x = 0$ and $x = 1$:

$$(5) \qquad u_x(0,t) = 0, \qquad u_x(1,t) = 0;$$

also, $u_x(x,t)$ is to be a continuous function of x on the closed interval $0 \leq x \leq 1$ for each fixed t ($t \geq t_0$). Then $u_x(0,t)$ is the same as the limit of $u_x(x,t)$ as $x \to 0$ from the interior of the interval, and similarly for $u_x(1,t)$; that is, conditions (5) then imply that

$$u_x(+0,t) = 0, \qquad u_x(1 - 0, t) = 0 \quad (t \geq t_0 > 0).$$

We noted that each function of the infinite sequence

$$u_0 = 1, \qquad u_n = \exp(-n^2\pi^2 t) \cos n\pi x \quad (n = 1, 2, \ldots)$$

satisfies the homogeneous equations (4) and (5). Those functions and their partial derivatives are continuous everywhere. According to Theorem 2 the function

$$(6) \qquad u(x,t) = c_0 + \sum_{n=1}^{\infty} c_n \exp(-n^2\pi^2 t) \cos n\pi x$$

also satisfies equations (4) and (5) if the constant coefficients c_n in the infinite series can be restricted so that the series is differentiable once with respect to t and twice with respect to x and so that u_x is continuous in x ($0 \leq x \leq 1$). We shall show why those conditions are satisfied when the sequence of constants

c_n is bounded, that is, when a number N exists such that

$$(7) \qquad\qquad |c_n| \leqq N \qquad (n = 0, 1, 2, \ldots).$$

The terms $c_n u_n$ of series (6) satisfy the condition

$$(8) \qquad |c_n \exp(-n^2\pi^2 t) \cos n\pi x| \leqq N \exp(-n^2\pi^2 t_0) \quad (t \geqq t_0)$$

and the ratio test shows the convergence of the series whose terms are the constants $N \exp(-n^2\pi^2 t_0)$ when $t_0 > 0$. Series (6) is therefore absolutely convergent by the comparison test and also uniformly convergent with respect to x and t according to the Weierstrass test. In like manner we see that

$$\left| \frac{\partial}{\partial x} (c_n u_n) \right| \leqq N\pi n \exp(-n^2\pi^2 t_0).$$

The series of constants $n^k \exp(-n^2\pi^2 t_0)$, where k is any integer, converges according to the ratio test. Hence the series whose terms are $(c_n u_n)_x$ is uniformly convergent with respect to x, in fact, with respect to x and t together. Series (6) is therefore differentiable with respect to x, and the sum $u_x(x,t)$ of that derived series is continuous with respect to x for all x when $t \geqq t_0$.

The uniform convergence of the series of derivatives $(c_n u_n)_{xx}$, which are the same as $(c_n u_n)_t$, is also evident. Thus series (6) is differentiable twice with respect to x and once with respect to t. Its sum u therefore satisfies conditions (4) and (5), and it involves an infinite set of constants c_n, arbitrary except for the boundedness condition (7).

We shall see later on that those constants can be determined so that u will satisfy a nonhomogeneous initial condition

$$(9) \qquad\qquad u(x,0) = f(x) \qquad (0 < x < 1).$$

Then, physically, u will represent temperatures in a slab $0 \leqq x \leqq 1$ with insulated faces, in view of conditions (5), and with initial temperatures given by condition (9). Note that condition (9) is satisfied by the function (6) if c_n can be found such that f has the representation

$$(10) \qquad\qquad f(x) = c_0 + \sum_{n=1}^{\infty} c_n \cos n\pi x \qquad (0 < x < 1).$$

PROBLEMS

1. Verify that each of the products $u_n = \cos nx \sinh ny$, where $n = 1$, 2, . . . , as well as the function $u_0 = y$ satisfies Laplace's equation $u_{xx} + u_{yy} = 0$ and the three boundary conditions

$$u_x(0,y) = u_x(\pi,y) = u(x,0) = 0.$$

Then apply Theorem 1 to show that the linear combination

$$u = c_0 y + \sum_{n=1}^{m} c_n \cos nx \sinh ny$$

also satisfies that differential equation and those boundary conditions. Note the values of the coefficients c_n for which that linear combination also satisfies the nonhomogeneous boundary condition

$$u(x,2) = 4 + 3 \cos x - \cos 2x.$$

Ans. $c_0 = 2$, $c_1 = 3/\sinh 2$, $c_2 = -1/\sinh 4$, $c_n = 0$ when $n > 2$.

2. Show that each of the functions

$$\exp\left[-(n - \tfrac{1}{2})^2 \pi^2 t\right] \sin\left[(n - \tfrac{1}{2})\pi x\right] \quad (n = 1, 2, \ldots)$$

satisfies the heat equation

$$u_t(x,t) = u_{xx}(x,t)$$

and the boundary conditions

$$u(0,t) = 0, \qquad u_x(1,t) = 0.$$

Then apply Theorem 1 to show that the function

$$u(x,t) = \sum_{n=1}^{m} c_n \exp\left[-(n - \tfrac{1}{2})^2 \pi^2 t\right] \sin\left[(n - \tfrac{1}{2})\pi x\right]$$

also satisfies those three equations. Note the values of the constants c_n such that this function also satisfies the nonhomogeneous boundary condition

$$u(x,0) = 2 \sin \frac{\pi x}{2} - \sin \frac{5\pi x}{2}.$$

Ans. $c_1 = 2$, $c_2 = 0$, $c_3 = -1$, $c_n = 0$ when $n > 3$.

3. Verify that each of the functions

$$\sin mx \cos ny \exp \left(-z \sqrt{m^2 + n^2}\right)$$
$$(m = 1, 2, \ldots ; n = 0, 1, 2, \ldots)$$

satisfies Laplace's partial differential equation

$$u_{xx}(x,y,z) + u_{yy}(x,y,z) + u_{zz}(x,y,z) = 0$$

and the boundary conditions

$$u(0,y,z) = u(\pi,y,z) = u_y(x,0,z) = u_y(x,\pi,z) = 0.$$

Use superposition to obtain a function that satisfies not only Laplace's equation and those boundary conditions, but also the condition

$$u_z(x,y,0) = (-6 + 5 \cos 4y) \sin 3x.$$

$$Ans. \ u = (2e^{-3z} - e^{-5z} \cos 4y) \sin 3x.$$

4. Let u and v denote two functions that satisfy the heat equation in x and t; that is, $u_t = ku_{xx}$ and $v_t = kv_{xx}$, where k may be a function of x and t. Multiply the members of those two equations by constants c_1 and c_2, respectively, and add to show that the linear combination $c_1u + c_2v$ also satisfies the equation. This illustrates a variation in the proof of Theorem 1.

5. If an operator L has the two properties

$$L(u_1 + u_2) = Lu_1 + Lu_2, \qquad L(cu) = cLu,$$

for all functions u_1, u_2, and u and for every constant c, show that L is linear; that is, show that it has the property (2), Sec. 11.

6. If u_1 and u_2 satisfy a linear nonhomogeneous differential equation $Lu = f$, where f is a function of the independent variables only, prove that the linear combination $c_1u_1 + c_2u_2$ fails to satisfy that equation if $c_1 + c_2 \neq 1$.

7. Note that each of the functions $y_1 = x^{-1}$ and $y_2 = (x + 1)^{-1}$ satisfies the nonlinear differential equation $y' + y^2 = 0$. If c is a constant $(c \neq 0, \ c \neq 1)$, show that neither cy_1 nor cy_2 satisfies that equation. Also show that the function $y_1 + y_2$ does not satisfy it.

8. Use special cases of linear operators, such as $Lu = xu$ and $Mu = \partial u/\partial x$, to show that the operators LM and ML are not always the same.

9. If c_n is a bounded sequence of constants and $y_0 > 0$, prove that the function

$$u(x,y) = \sum_{n=1}^{\infty} c_n e^{-ny} \sin nx \qquad (y \geq y_0)$$

is twice differentiable with respect to x and y and satisfies Laplace's equation in the domain $y > y_0$.

10. If $n^4|c_n| \leqq N$ $(n = 1, 2, \ldots)$, where N is a constant, prove that the function

$$y(x,t) = \sum_{n=1}^{\infty} c_n \sin nx \cos nt$$

satisfies the wave equation $y_{tt} = y_{xx}$ for all x and t.

15. Integrals of Solutions. Another type of extension of the concept of linear combinations of functions satisfying linear homogeneous conditions is illustrated by the following example. Here superposition consists of integration with respect to a parameter α instead of summation with respect to a parameter n.

The functions of the set exp $(-\alpha y)$ sin αx, where the parameter α is independent of x and y, satisfy Laplace's equation

$$(1) \qquad\qquad u_{xx}(x,y) + u_{yy}(x,y) = 0$$

and the boundary condition

$$(2) \qquad\qquad u(0,y) = 0,$$

Those functions are bounded in the domain $x > 0$, $y > 0$ if $\alpha \geqq 0$. We can show that their combination of the type

$$(3) \qquad u(x,y) = \int_0^{\infty} g(\alpha)e^{-\alpha y} \sin \alpha x \, d\alpha \qquad (x > 0, y > 0$$

also represents a solution of the homogeneous conditions (1) and (2), bounded in the domain $x > 0$, $y > 0$ for each function $g(\alpha)$ that is bounded and continuous on the semi-infinite interval $\alpha \geqq 0$ and absolutely integrable over that interval.

The function (3) satisfies the boundary condition

$$(4) \qquad u(x,0) = \int_0^{\infty} g(\alpha) \sin \alpha x \, d\alpha \qquad\qquad (x > 0)$$

on the positive x axis.   Our theory of Fourier integrals will show how g can be determined, even when the above conditions on g are relaxed somewhat, so that the integral in equation (4) will represent a given function $f(x)$.

To show that equation (3) represents a bounded continuous function u that satisfies equations (1) and (2), we use tests of

improper integrals that are analogous to those for infinite series.[1]
The integral in equation (3) converges absolutely and uniformly
with respect to x and y because

$$(5) \qquad\qquad |g(\alpha)e^{-\alpha y} \sin \alpha x| \leq |g(\alpha)| \qquad (x \geq 0, y \geq 0)$$

and g is independent of x and y and absolutely integrable from
zero to infinity with respect to α.   Also

$$(6) \qquad |u(x,y)| \leq \int_0^\infty |g(\alpha)e^{-\alpha y} \sin \alpha x| \, d\alpha \leq \int_0^\infty |g(\alpha)| \, d\alpha,$$

so that u is bounded; it is also a continuous function of x and
y $(x \geq 0, y \geq 0)$ because of the uniform convergence of the
integral and the continuity of the integrand.   Clearly $u = 0$
when $x = 0$.

When $y > 0$ it is true that

$$(7) \qquad \frac{\partial u}{\partial x} = \frac{\partial}{\partial x} \int_0^\infty g(\alpha)e^{-\alpha y} \sin \alpha x \, d\alpha = \int_0^\infty \frac{\partial}{\partial x}[g(\alpha)e^{-\alpha y} \sin \alpha x] \, d\alpha;$$

for if $|g(\alpha)| \leq g_0$ and $y \geq y_0$, where y_0 is some small positive
number, then the absolute value of the integrand of the integral
on the right does not exceed $g_0\alpha \exp(-\alpha y_0)$, which is independent
of x and y and integrable from $\alpha = 0$ to $\alpha = \infty$.   Hence that
integral is uniformly convergent; thus integral (3) is differentiable
with respect to x, and similarly for the other derivatives involved
in the laplacian operator $\nabla^2 = \partial^2/\partial x^2 + \partial^2/\partial y^2$.   Therefore

$$(8) \qquad\qquad \nabla^2 u = \int_0^\infty g(\alpha)\nabla^2[e^{-\alpha y} \sin \alpha x] \, d\alpha = 0 \quad (x > 0, y > 0).$$

16. Separation of Variables.   Let us find a formula for the
transverse displacements $y(x,t)$ of a string stretched between the
points $(0,0)$ and $(c,0)$ if the string is initially displaced into a
position $y = f(x)$ and released at rest from that position.

At this stage some of our steps in solving this problem must be
formal, or manipulative.   Their validity cannot be established
until we have developed further theory.

We assume that no external forces act along the string.   Then
the function y satisfies the wave equation (Sec. 3)

$$(1) \qquad\qquad y_{tt} = a^2 y_{xx}(x,t) \qquad (0 < x < c, t > 0).$$

[1] Kaplan, *op. cit.*, pp. 372ff.

It must also satisfy these boundary conditions:

(2) $\qquad y(0,t) = 0, \qquad y(c,t) = 0, \qquad y_t(x,0) = 0,$

(3) $\qquad\qquad\qquad y(x,0) = f(x) \qquad\qquad (0 \leqq x \leqq c),$

where the prescribed displacement function f is continuous on the interval $0 \leqq x \leqq c$ and $f(0) = f(c) = 0$.

In order to find an extensive set of particular solutions *of all homogeneous conditions* (1) and (2) in the above boundary value problem, using ordinary differential equations, we first determine those functions of type

(4) $\qquad\qquad\qquad Y(x,t) = X(x)T(t)$

which satisfy those conditions. Note that X is a function of x alone and T a function of t alone.

If Y satisfies equation (1), then

$$X(x)T''(t) = a^2 X''(x)T(t).$$

We divide by $a^2 XT$ here to separate the variables:

(5) $\qquad\qquad\qquad \dfrac{X''(x)}{X(x)} = \dfrac{T''(t)}{a^2 T(t)}.$

Since the member on the left is a function of x alone, it cannot vary with t. However, it is equal to a function of t alone and so it cannot vary with x. Hence both members must have some constant value, which we write as $-\lambda$, in common; that is,

(6) $\qquad X''(x) = -\lambda X(x). \qquad T''(t) = -\lambda a^2 T(t).$

If Y is to satisfy the first of conditions (2), then $X(0)T(t)$ must vanish for all t $(t > 0)$. The case $T(t) = 0$ for all t is trivial since the function $Y = 0$ always satisfies linear homogeneous equations; hence $X(0) = 0$. Likewise, the last two of conditions (2) are satisfied by Y if $X(c) = 0$ and $T'(0) = 0$.

Thus Y satisfies conditions (1) and (2) when X and T satisfy these two homogeneous problems:

(7) $\qquad X''(x) + \lambda X(x) = 0, \qquad X(0) = 0, \qquad X(c) = 0,$

(8) $\qquad\qquad T''(t) + \lambda a^2 T(t) = 0, \qquad T'(0) = 0,$

where the **parameter** λ has the same value in both problems. To find nontrivial solutions of this pair of problems we first note that problem (8) has only one boundary condition and therefore many solutions for each value of λ. Since problem (7) has two boundary conditions, it may have nontrivial solutions for exceptional values of λ. As we shall see in the next chapter, problem (7) is a special case of the Sturm-Liouville problem, and λ must be real-valued if the problem is to have a nontrivial solution.

If $\lambda = 0$, the differential equation in problem (7) becomes $X''(x) = 0$ and its general solution is $X = Ax + B$. Since $B = 0$ if $X(0) = 0$, then $X(c) = Ac = 0$ only if $A = 0$; therefore this problem has just the trivial solution $X(x) = 0$ when $\lambda = 0$.

If $\lambda > 0$ the general solution of $X'' + \lambda X = 0$ is

$$X(x) = C_1 \sin x \sqrt{\lambda} + C_2 \cos x \sqrt{\lambda}.$$

Then $X(0) = 0$ if $C_2 = 0$ and $X(c) = 0$ if $\sin c \sqrt{\lambda} = 0$; that is,

$$(9) \qquad\qquad \sqrt{\lambda} = \frac{n\pi}{c} \qquad\qquad (n = 1, 2, \ldots).$$

Then, except for a constant factor,

$$(10) \qquad\qquad X = \sin \frac{n\pi x}{c} \qquad\qquad (n = 1, 2, \ldots).$$

The numbers $\lambda = n^2\pi^2/c^2$ for which problem (7) has nontrivial solutions are called *eigenvalues* of that problem, and functions (10) are the corresponding *eigenfunctions*.

When $\lambda < 0$, let us write $\lambda = -\mu^2$, where μ is real. Then

$$X = C_3 \sinh \mu x$$

is the solution of $X'' - \mu^2 X = 0$ that satisfies the condition $X(0) = 0$. Since $\sinh \mu c \neq 0$, $C_3 = 0$ if $X(c) = 0$. Thus problem (7) has no negative eigenvalues.

When $\lambda = n^2\pi^2/c^2$, problem (7) is a distinct problem for each different positive integer n. For a fixed integer n, it has the solution (10), and problem (8) becomes

$$T''(t) + n^2\pi^2 a^2 c^{-2} T(t) = 0, \qquad T'(0) = 0;$$

thus except for a constant factor,

$$T = \cos \frac{n\pi a t}{c}.$$

Therefore each function of the infinite set

$$(11) \qquad Y_n(x,t) = \sin \frac{n\pi x}{c} \cos \frac{n\pi a t}{c} \qquad (n = 1, 2, \ldots)$$

satisfies all the homogeneous conditions (1) and (2).

The procedure here illustrates a fundamental method of obtaining particular solutions of homogeneous conditions in boundary value problems, the method of *separation of variables*.

A linear combination of a finite number of the functions (11) also satisfies the homogeneous conditions (Theorem 1). But when $t = 0$, it reduces to a linear combination of a finite number of the functions $\sin (n\pi x/c)$. Thus it will not satisfy the non-homogeneous condition (3) unless $f(x)$ has that special character.

Note that a sum of functions of type $X(x)T(t)$ is not generally a function of that type. The sum $t^2 + xt$, for example, is not of that type.

According to Theorem 2, the function

$$(12) \qquad y(x,t) = \sum_{n=1}^{\infty} b_n \sin \frac{n\pi x}{c} \cos \frac{n\pi a t}{c}$$

also satisfies conditions (1) and (2), provided the coefficients b_n can be restricted adequately. That function will satisfy the remaining condition (3) if f can be represented in the form

$$(13) \qquad f(x) = \sum_{n=1}^{\infty} b_n \sin \frac{n\pi x}{c} \qquad (0 \leqq x \leqq c).$$

In the following section we shall indicate why the coefficients b_n should have the values

$$(14) \qquad b_n = \frac{2}{c} \int_0^c f(x) \sin \frac{n\pi x}{c} \, dx \qquad (n = 1, 2, \ldots)$$

if the representation (13) is to be valid.

The formal solution of our boundary value problem in the displacements of the string is formula (12), in which the coefficients have the values (14).

17. The Fourier Sine Series.   We recall that

$$2 \sin \frac{m\pi x}{c} \sin \frac{n\pi x}{c} = \cos \frac{(m-n)\pi x}{c} - \cos \frac{(m+n)\pi x}{c},$$

$$2 \sin^2 \frac{n\pi x}{c} = 1 - \cos \frac{2n\pi x}{c};$$

hence, when m and n are integers,

$$(1) \qquad \int_0^c \sin \frac{m\pi x}{c} \sin \frac{n\pi x}{c} \, dx = \begin{cases} 0 & \text{if } m \neq n \\ \dfrac{c}{2} & \text{if } m = n. \end{cases}$$

The set of functions $\sin (n\pi x/c)$ $(n = 1, 2, \ldots)$ is therefore *orthogonal* on the interval $0 < x < c$; that is, the integral, over that interval, of the product of any two distinct functions of the set is zero.

In Sec. 16 we needed to determine the coefficients b_n so that a series of those sine functions would converge to a prescribed function $f(x)$ on the interval $0 \leqq x \leqq c$. *Assuming* that such an expansion is possible, namely,

$$(2) \quad f(x) = b_1 \sin \frac{\pi x}{c} + b_2 \sin \frac{2\pi x}{c} + \cdots + b_n \sin \frac{n\pi x}{c} + \cdots,$$

and that the series can be integrated term by term after being multiplied by $\sin (n\pi x/c)$, we can use the orthogonality property (1) to find a formula for b_n.

Let all terms in equation (2) be multiplied by $\sin (n\pi x/c)$ and integrated from $x = 0$ to $x = c$. The first term on the right becomes

$$b_1 \int_0^c \sin \frac{\pi x}{c} \sin \frac{n\pi x}{c} \, dx.$$

This is zero unless $n = 1$, according to the orthogonality property. Likewise all other terms on the right, except the nth one, become zero.   In view of property (1), then

$$\int_0^c f(x) \sin \frac{n\pi x}{c} \, dx = b_n \int_0^c \sin^2 \frac{n\pi x}{c} \, dx = \frac{1}{2} cb_n.$$

Hence the coefficients in the representation (2) have the values

$$(3) \qquad b_n = \frac{2}{c} \int_0^c f(x) \sin \frac{n\pi x}{c} \, dx \qquad (n = 1, 2, \ldots).$$

The sine series (2) with coefficients (3) is called the *Fourier sine series* corresponding to the function f, on the interval $0 < x < c$. Using the tilde symbol $\sim$ to indicate correspondence, we can write

$$(4) \qquad f(x) \sim \frac{2}{c} \sum_{n=1}^{\infty} \sin \frac{n\pi x}{c} \int_0^c f(\xi) \sin \frac{n\pi \xi}{c}\, d\xi \quad (0 < x < c).$$

In Chap. 4 we shall establish general conditions on f under which this series converges to the values $f(x)$ so that the correspondence (4) becomes an equality.

18. A Plucked String. As a special case of the boundary value problem considered in Sec. 16, let the string be stretched between

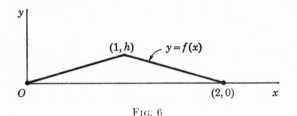

Fig. 6

the points $(0,0)$ and $(2,0)$ and suppose its mid-point is raised to a height h above the x axis. The string is then released from rest in that broken-line position (Fig. 6).

In this case the function f, which describes the initial position of the string, is given by the formula

$$(1) \qquad f(x) = \begin{cases} hx & \text{when } 0 \leqq x \leqq 1, \\ -hx + 2h & \text{when } 1 \leqq x \leqq 2. \end{cases}$$

The coefficients b_n in the Fourier sine series corresponding to that function, on the interval $0 \leqq x \leqq 2$, can be written

$$b_n = \int_0^2 f(x) \sin \frac{n\pi x}{2}\, dx$$

$$= h \int_0^1 x \sin \frac{n\pi x}{2}\, dx + h \int_1^2 (-x + 2) \sin \frac{n\pi x}{2}\, dx.$$

After integrating and simplifying, we find that

$$(2) \qquad b_n = \frac{8h}{\pi^2 n^2} \sin \frac{n\pi}{2}.$$

According to our formal solution (12), Sec. 16, then,

$$(3) \qquad y(x,t) = \frac{8h}{\pi^2} \sum_{n=1}^{\infty} \frac{1}{n^2} \sin \frac{n\pi}{2} \sin \frac{n\pi x}{2} \cos \frac{n\pi at}{2}.$$

But $\sin (n\pi/2) = 0$ when n is even. When n is odd, we may write $n = 2m - 1$ where $m = 1, 2, \ldots$; then

$$\sin \frac{n\pi}{2} = \sin \left(m - \frac{1}{2} \right) \pi = - \cos m\pi \sin \frac{\pi}{2} = (-1)^{m+1}.$$

Thus formula (3) for the displacements takes the form

$$(4) \quad y(x,t) = \frac{8h}{\pi^2} \sum_{m=1}^{\infty} \frac{(-1)^{m+1}}{(2m-1)^2} \sin \frac{(2m-1)\pi x}{2} \cos \frac{(2m-1)\pi at}{2}.$$

The infinite series in formulas (3) and (4) is not differentiable twice with respect to x or t. But in Sec. 56 we shall show how the series can be represented in a form simple enough to see that the sum $y(x,t)$ of the series is a solution of the boundary value problem.

PROBLEMS

1. It was pointed out in Sec. 12 that the homogeneous differential equation $u_t = u_{xx}$ and boundary conditions $u_x(0,t) = 0$, $u_x(1,t) = 0$ are all satisfied by these functions of type $T(t)X(x)$:

$$u_0 = 1, \qquad u_n = \exp (-n^2\pi^2 t) \cos n\pi x \qquad (n = 1, 2, \ldots).$$

Use the method of separation of variables (Sec. 16) to derive that set of particular solutions. Note that the solution $u_0 = 1$ corresponds to the eigenvalue $\lambda = 0$ for the problem $X'' + \lambda X = 0$, $X'(0) = X'(1) = 0$.

2. Use the method of separation of variables to obtain the set of functions $\exp [-(n - \frac{1}{2})^2\pi^2 t] \sin [(n - \frac{1}{2})\pi x]$ used in Problem 2, Sec. 14.

3. In Problem 1, Sec. 14, it was pointed out that each function of the set

$$u_0 = y, \qquad u_n = \cos nx \sinh ny \qquad (n = 1, 2, \ldots)$$

satisfies Laplace's equation and the homogeneous boundary conditions

$$u_x(0,y) = u_x(\pi,y) = u(x,0) = 0.$$

Use the method of separation of variables to derive that set.

4. Show that the Fourier sine series corresponding to the function $f(x) = 3 \sin 4\pi x$ on the interval $(0 < x < 1)$ reduces to $3 \sin 4\pi x$, the function itself.

5. Show that the Fourier sine series corresponding to the function $f(x) = 1$ on the interval $0 < x < \pi$ is

$$\frac{4}{\pi} \sum_{n=1}^{\infty} \frac{\sin(2n-1)x}{2n-1} = \frac{4}{\pi} \left(\sin x + \frac{1}{3} \sin 3x + \cdots \right).$$

6. Write the Fourier sine series corresponding to the function $f(x) = x$ on the interval $0 < x < 1$.

$$Ans. \; \frac{2}{\pi} \sum_{n=1}^{\infty} \frac{(-1)^{n+1}}{n} \sin n\pi x.$$

19. Ordinary Differential Equations. In Sec. 16 we found that the homogeneous boundary value problem

$$(1) \qquad X''(x) + \lambda X = 0, \qquad X(0) = 0, \qquad X(c) = 0$$

has many solutions $X = C_1 \sin (n\pi x/c)$, one for each value of C_1, in case the coefficient λ has the value $n^2\pi^2/c^2$, where n is some positive integer. When $\lambda \neq n^2\pi^2/c^2$, the problem has the unique solution $X(x) = 0$. This illustrates the fact that uniqueness of solutions of boundary value problems may depend on the coefficients in the differential equation and the kind of boundary conditions, not solely on the order of the equation and the number of boundary conditions. This is also true of the existence of solutions as we shall point out below.

The theory of ordinary differential equations ensures the existence and uniqueness of solutions of certain types of *initial value problems*, problems in which all boundary data are given at one point. We state, without proof, a theorem for linear equations of the second order:

Let A, B, and C denote continuous functions of x on an interval $a \leq x \leq b$. If x_0 is a fixed point on that interval and y_0 and y_0' are prescribed constants, then there is one and only one function y, continuous together with its derivative y' on that interval, which satisfies the differential equation

$$(2) \qquad y''(x) + A(x)y'(x) + B(x)y(x) = C(x) \qquad (a < x < b)$$

and the two initial conditions

$$(3) \qquad\qquad y(x_0) = y_0, \qquad y'(x_0) = y_0'.$$

Proofs of this and corresponding theorems for equations of other orders, by methods of successive approximations, will be found in books on the theory of differential equations.[1]   According to equation (2), $y'' = C - Ay' - By$; thus y'' is also continuous.   That the equation has a general solution involving two arbitrary constants follows from the fact that arbitrary values can be assigned to the constants y_0 and y_0'.

As an example, the equation

$$(4) \qquad\qquad x''(t) + k^2 x(t) = 0,$$

where k is a constant and $k \neq 0$, has a solution

$$(5) \qquad\qquad x = C_1 \sin kt + C_2 \cos kt$$

whenever C_1 and C_2 are constants.   On each interval of the t axis that contains the point $t = 0$ the function (5) is *the* solution of equation (4) that satisfies the conditions $x(0) = C_2$ and $x'(0) = kC_1$.

Suppose x is to satisfy the two boundary conditions

$$(6) \qquad\qquad x(0) = 0, \qquad x(1) = 1$$

along with equation (4) and the requirement that x and x' be continuous ($0 \leq t \leq 1$).   Then if $x(0) = 0$ it is necessary that

$$x(t) = C_1 \sin kt.$$

If $x(1) = 1$, a value of C_1 must exist such that

$$(7) \qquad\qquad C_1 \sin k = 1.$$

This equation determines C_1 provided $\sin k \neq 0$; thus

$$(8) \qquad\qquad x = \frac{\sin kt}{\sin k} \qquad (k \neq \pm n\pi,\ n = 1, 2, \ldots).$$

But when $k = \pm n\pi$, the two-point boundary value problem consisting of equations (4) and (6) *has no solution* because there is no value of C_1, or $x'(0)/k$, that satisfies equation (7).

It is interesting to note a physical interpretation of the case

[1] See p. 73 of the book by Ince listed in the Bibliography.

$k = \pi$. Since $x'' = -\pi^2 x$, the function x represents the displacement of a unit mass along the x axis under a force $-\pi^2 x$ proportional to the displacement from the origin $x = 0$. The mass is initially at the origin; it is required to have the displacement $x = 1$ at time $t = 1$. But the number π is the natural frequency k of this system, and as a consequence the mass returns to the origin at the instants $t = 1, 2, \ldots$ regardless of its initial velocity. Therefore $x(1) = 0$, and the condition $x(1) = 1$ cannot be satisfied.

20. General Solutions in Partial Differential Equations. Some boundary value problems in partial differential equations can be solved by a method corresponding to the one usually used to solve problems in ordinary differential equations, the method of using a general solution of the differential equation.

Example 1.   Solve the boundary value problem

$$(1) \qquad u_{xx}(x,y) = 0, \qquad u(0,y) = y^2, \qquad u(1,y) = 1.$$

Successive integrations of the equation $u_{xx} = 0$ with respect to x, with y kept fixed, give the equations $u_x = f(y)$ and

$$(2) \qquad u = xf(y) + g(y),$$

where f and g are arbitrary functions. The boundary conditions in problem (1) require that

$$g(y) = y^2, \qquad f(y) + g(y) = 1;$$

thus $f(y) = 1 - y^2$ and the solution of the problem is

$$(3) \qquad u = x(1 - y^2) + y^2.$$

Example 2.   Solve the wave equation

$$(4) \qquad y_{tt}(x,t) = a^2 y_{xx}(x,t)$$

in the domain $-\infty < x < \infty$, $t > 0$, subject to the boundary conditions

$$(5) \qquad y(x,0) = f(x), \qquad y_t(x,0) = 0 \quad (-\infty < x < \infty),$$

in terms of the constant a and the function f.

The differential equation (4) can be simplified by introducing new independent variables r and s such that

$$r = x + at, \qquad s = x - at.$$

According to the chain rule for differentiation,

$$y_t = y_r r_t + y_s s_t = a y_r - a y_s.$$

Continuing with that rule we find that

$$y_{tt} = a^2(y_{rr} - 2y_{rs} + y_{ss}), \qquad y_{xx} = y_{rr} + 2y_{rs} + y_{ss}.$$

Equation (4) therefore reduces to the form

$$y_{rs}(r,s) = 0$$

which can be solved by successive integrations to give $y_r = g'(r)$ and

$$y = g(r) + h(s),$$

where g and h are arbitrary differentiable functions.

A general solution of the wave equation (4) is therefore

(6) $$y = g(x + at) + h(x - at).$$

In this example the boundary conditions are simple enough that we can determine the functions g and h, because the function (6) satisfies conditions (5) when

$$g(x) + h(x) = f(x) \qquad \text{and} \qquad ag'(x) - ah'(x) = 0.$$

Thus $g(x) - h(x) = c$, a constant, and $2h + c = f$; that is, $2h = f - c$ and $2g = f + c$ so that

(7) $$y = \tfrac{1}{2}[f(x + at) + f(x - at)].$$

The solution (7) of the boundary value problem consisting of equations (4) and (5) is easily verified, assuming that $f'(x)$ and $f''(x)$ exist for all x.

The method illustrated in the two examples here has severe limitations. The general solutions (2) and (6), solutions involving arbitrary functions, were obtained by successive integrations, a procedure that applies to relatively few types of partial differential equations. But even in the exceptional cases where such general solutions can be found the determination of the arbitrary functions directly from the boundary conditions is often too difficult.

21. Superposition. Other Methods. We have stressed a process based on superposition for solving linear boundary value problems in partial differential equations. The process, illustrated in Sec. 16, consists first of finding particular solutions of all homogeneous conditions in the problem by the method of separation of variables. A generalized linear combination, or superposition, of those solutions is then sought which will satisfy the nonhomogeneous conditions too.

This method of *separation of variables and superposition,* sometimes called the *Fourier method,* is a classical and powerful one. The treatment of its theory and applications is the objective of this book. Limitations of the method will be noted also.

There are other important methods of solving linear boundary value problems. The procedures of using Laplace transforms, Fourier transforms, or other integral transforms, all included in the subject of operational mathematics, are especially effective.[1] The classical method of conformal mapping in the theory of functions of a complex variable applies to a prominent class of problems involving Laplace's equation in two dimensions.[2] There are still other ways of reducing or solving such problems, including applications of Green's functions and numerical or computational methods.

Even when a problem yields to more than one method, however, different methods sometimes produce different forms of the solution, and each form may have its own desirable features. On the other hand, some problems require successive applications of two or more methods. Others, including some linear problems of fairly simple types, have defied all known exact methods. The development of new methods is an activity in present-day mathematical research.

PROBLEMS

Use general solutions of the partial differential equations to solve the boundary value problems in Problems 1 to 4.

1. $u_{xx}(x,y) = 6xy$, $u(0,y) = y$, $u_x(1,y) = 0$.
$$Ans.\ u = (x^3 - 3x + 1)y.$$

2. $u_{xy}(x,y) = 2x$, $u(0,y) = 0$, $u(x,0) = x^2$. $Ans.\ u = x^2(1 + y)$.

3. $y_{tt}(x,t) = a^2 y_{xx}(x,t)$, $y(x,0) = 0$, $y_t(x,0) = (1 + x^2)^{-1}$.
$$Ans.\ 2ay = \arctan (x + at) - \arctan (x - at).$$

[1] See the author's "Operational Mathematics," listed in the Bibliography.
[2] Churchill, R. V., "Complex Variables and Applications," 2d ed., 1960.

4. $y_{tt}(x,t) = a^2 y_{xx}(x,t)$, $y(x,0) = 0$, $y_t(x,0) = g(x)$.

$$\text{Ans. } y = (2a)^{-1} \int_{x-at}^{x+at} g(\xi)\, d\xi.$$

5. Use superposition of the solution (7), Sec. 20, and the solution of Problem 4 to write the solution

$$y = \tfrac{1}{2}[f(x + at) + f(x - at)] + \frac{1}{2a} \int_{x-at}^{x+at} g(\xi)\, d\xi$$

of the boundary value problem

$$y_{tt}(x,t) = a^2 y_{xx}(x,t), \qquad y(x,0) = f(x), \qquad y_t(x,0) = g(x).$$

6. In Example 2 of Sec. 20, $y(x,t)$ represents transverse displacements of a stretched string of infinite length, initially released at rest from a position $y = f(x)$ $(-\infty < x < \infty)$. From the solution $y = \tfrac{1}{2}f(x + at) + \tfrac{1}{2}f(x - at)$ show why the instantaneous position of the string at time t can be described as the curve obtained by adding ordinates of two curves, one obtained by translating the curve $y = \tfrac{1}{2}f(x)$ to the left through the distance at, the other by translating it to the right through the same distance. As t varies, the curve $y = \tfrac{1}{2}f(x)$ moves as a wave with velocity a. Show graphically some instantaneous positions of the string when $f(x)$ is zero except on a small interval about the origin.

7. Derive the general solution

$$u = x + e^{-y}g(x) + h(y)$$

of the partial differential equation $u_{xy} + u_x = 1$.

8. The boundary value problem

$$
\begin{aligned}
y_{tt}(x,t) &= a^2 y_{xx}(x,t) & (x > 0, t > 0),\\
y(x,0) = e^{-x}, &\quad y_t(x,0) = 0 & (x > 0),\\
y_x(0,t) = y(0,t), &\quad \lim_{x \to \infty} y(x,t) = 0 & (t > 0)
\end{aligned}
$$

can be solved with the aid of elementary Laplace transforms. Its solution is given by the formula

$$
y = \begin{cases}
e^{-x} \cosh at & \text{when } x \geqq at,\\
e^{-x} \cosh at + \sinh(x - at) + (x - at)\exp(x - at) & \\
& \text{when } x \leqq at.
\end{cases}
$$

Show that this formula is a special case of the general solution (6), Sec. 20, and hence that y does satisfy the wave equation in the domain $x > 0$, $y > 0$ except at points on the line $x = at$ where the derivatives of y do not exist. Observe that when $t = 0$ and $x > 0$, $y = e^{-x}$ here since $x > at$. Show that y satisfies the rest of the boundary conditions. [*Remark:* Even though the general solution of the partial differential equation here is known, this fairly simple boundary value problem is poorly adapted to the method (Sec. 20) of using the general solution.]

22. Historical Development. Mathematical sciences experienced a burst of activity following the invention of calculus by Newton (1642–1727) and Leibnitz (1646–1716). Among topics in mathematical physics that attracted the attention of great scientists during that period were boundary value problems in vibrations of strings stretched between fixed points and vibrations of bars or columns of air, associated with mathematical theories of musical vibrations. Early contributors to the theory of vibrating strings included the English mathematician Brook Taylor (1685–1731), the Swiss mathematicians Daniel Bernoulli (1700–1782) and L. Euler (1707–1783), and d'Alembert (French, 1717–1783).

By the 1750s, d'Alembert, Bernoulli, and Euler had advanced the theory of vibrating -strings to the stage where the partial differential equation $y_{tt} = a^2 y_{xx}$ was known and a solution of the boundary value problem had been found from the general solution of that equation. Also, the concept of fundamental modes of vibration led those men to notions of superposition of solutions, to a solution of the form (12) in Sec. 16 as a series of trigonometric functions and thus to the matter of representing an arbitrary function by a trigonometric series. Later on, Euler gave the formulas for the coefficients in the series. But the general concept of a function had not been clarified, and a lengthy controversy took place over the question of representing arbitrary functions on a finite interval by series of sine functions. This question of representation was first settled by Dirichlet about 70 years later.

The French mathematical physicist J. B. Fourier (1768–1830) presented many instructive examples of expansions in trigonometric series in connection with boundary value problems in the conduction of heat. His book "Théorie analytique de la chaleur," published in 1822, is a classic on the theory of heat conduction. He effectively illustrated the basic procedures of separation of variables and superposition, and his work did much toward arousing interest in trigonometric series representations.

But Fourier's contributions to the representation problem did not include conditions of validity of the representation; he was interested in applications and methods.   The German mathematician P. G. Lejeune Dirichlet (1805–1859) was the first to give such conditions.   It was in 1829 that Dirichlet firmly established general conditions on a function sufficient to ensure the convergence of its Fourier series to values of the function.[1]

Representation theory has been refined and greatly extended since Dirichlet's time.   It is still growing.

[1] For supplementary reading on history of Fourier series see the articles by R. E. Langer and E. B. Van Vleck listed in the Bibliography.

CHAPTER 3

ORTHOGONAL SETS OF FUNCTIONS

23. The Inner Product of Two Vectors. The concept of an orthogonal set of functions is a natural generalization of the concept of an orthogonal set of vectors, that is, a set of mutually perpendicular vectors. In fact, a function can be considered as a generalized vector so that fundamental properties of the set of functions are suggested by the analogous properties of the set of vectors. In the following discussion of vectors, the terminology and notation which apply to the generalized case will be used whenever it seems advantageous for the later generalizations.

Let either g or $g(r)$ denote a vector in ordinary three-dimensional space whose rectangular components are the three numbers $g(1)$, $g(2)$, and $g(3)$. It is the radius vector of the point having those three numbers as rectangular cartesian coordinates. The length of that vector, called its *norm*, will be written $\|g\|$; it is the nonnegative number such that

$$(1) \qquad \|g\|^2 = [g(1)]^2 + [g(2)]^2 + [g(3)]^2 = \sum_{r=1}^{3} [g(r)]^2.$$

If $\|g\| = 1$, g is a unit vector, also called a *normed* or *normalized* vector. If $\|g\| = 0$, then each of its components is zero, and conversely; g is then the *zero vector*.

The linear combination $c_1 g_1 + c_2 g_2$ of two vectors g_1 and g_2, where c_1 and c_2 are numbers, is the vector whose components are the three numbers $c_1 g_1(r) + c_2 g_2(r)$, $r = 1, 2, 3$; similarly for three or more vectors. Vector addition and subtraction, and multiplication by a number, are linear combinations. We note in particular that the norm of the vector $g_1 - g_2$,

$$(2) \qquad \|g_1 - g_2\| = \left\{ \sum_{r=1}^{3} [g_1(r) - g_2(r)]^2 \right\}^{\frac{1}{2}},$$

is the distance between the tips of the vectors g_1 and g_2.

The scalar product, or *inner product*, of g_1 and g_2 is denoted by the symbol (g_1,g_2); thus

$$(3) \qquad (g_1,g_2) = \sum_{r=1}^{3} g_1(r)g_2(r) = \|g_1\| \, \|g_2\| \cos \theta,$$

where θ is the angle between the two vectors, defined when neither is the zero vector. In case $\|g_2\| = 1$, then (g_1,g_2) is the projection of g_1 on the direction of g_2. The condition that two nonzero vectors be *orthogonal*, or perpendicular to each other, can be written

$$(4) \qquad (g_1,g_2) = 0, \qquad \text{or} \qquad \sum_{r=1}^{3} g_1(r)g_2(r) = 0.$$

In view of equations (1) and (3), the norm of g can be written

$$(5) \qquad \|g\| = (g,g)^{\frac{1}{2}}.$$

24. Orthonormal Sets of Vectors. Given an orthogonal set of three nonzero vectors g_n $(n = 1, 2, 3)$, a set of unit vectors ϕ_n having the same directions can be formed by dividing each vector g_n by its length. The components of ϕ_1, for instance, are $\phi_1(r) = \|g_1\|^{-1}g_1(r)$ $(r = 1, 2, 3)$. This set of mutually perpendicular unit vectors ϕ_n is called an *orthonormal set*. Such a set is described by means of inner products by writing

$$(1) \qquad (\phi_m,\phi_n) = \delta_{mn} \qquad (m, n = 1, 2, 3),$$

where δ_{mn} is *Kronecker's δ*:

$$\delta_{mn} = \begin{cases} 0 & \text{if } m \neq n, \\ 1 & \text{if } m = n. \end{cases}$$

Condition (1) therefore states that each vector of the set ϕ_1, ϕ_2, ϕ_3 is perpendicular to the other two and that each has unit length.

The symbol $\{\phi_n\}$ will be used to denote such orthonormal sets. A simple example is the set consisting of unit vectors along the three coordinate axes.

Every vector f in the space can be expressed as a linear combination of the vectors ϕ_1, ϕ_2, and ϕ_3. That is, three numbers c_1, c_2, and c_3 can be found for which

$$(2) \qquad f(r) = c_1\phi_1(r) + c_2\phi_2(r) + c_3\phi_3(r) \quad (r = 1, 2, 3),$$

when the components $f(1)$, $f(2)$, and $f(3)$ are given. To find the number c_1 in a simple way, consider equation (2) as a vector equation and take the inner product of both its members by ϕ_1. This gives

$$(f,\phi_1) = c_1(\phi_1,\phi_1) + c_2(\phi_2,\phi_1) + c_3(\phi_3,\phi_1) = c_1$$

since $(\phi_1,\phi_1) = 1$ and $(\phi_2,\phi_1) = (\phi_3,\phi_1) = 0$, according to condition (1). Similarly c_2 and c_3 are found by taking the inner product of the members of equation (2) by ϕ_2 and ϕ_3, respectively. The coefficients are therefore

$$(3) \qquad c_n = (f,\phi_n) = \sum_{r=1}^{3} f(r)\phi_n(r) \qquad (n = 1, 2, 3);$$

c_n is the projection of f on ϕ_n. The representation (2) can then be written

$$(4) \qquad f(r) = (f,\phi_1)\phi_1(r) + (f,\phi_2)\phi_2(r) + (f,\phi_3)\phi_3(r)$$
$$= \sum_{n=1}^{3} (f,\phi_n)\phi_n(r).$$

The representation (2), or (4), may be called an expansion of the arbitrary vector f in a finite series of the orthonormal reference vectors. Those orthogonal reference vectors were assumed to be normalized only as a matter of convenience in order to obtain the simple formulas (3) for the coefficients in the expansion.

The definitions and results just given can be extended immediately to vectors in a space of k dimensions. In this case the index r, which indicates the component, has values from 1 to k instead of 1 to 3. The definition of the inner product of two vectors g_1 and g_2 in this space, for instance, becomes

$$(5) \qquad (g_1,g_2) = \sum_{r=1}^{k} g_1(r)g_2(r).$$

A generalization of another sort is also possible. The units of length on the rectangular coordinate axes, with respect to which the components of vectors are measured, may vary from one axis to another. In such a case the scalar product of two vectors g_1 and g_2 in three-dimensional space has the form

$$(g_1,g_2) = \sum_{r=1}^{3} p(r)g_1(r)g_2(r).$$

The weight numbers $p(1)$, $p(2)$, and $p(3)$ here depend upon the units of length used along the three axes.

25. Functions as Vectors. Orthogonality.   A vector $g(r)$ in three dimensions was described above by the numbers $g(1)$, $g(2)$, and $g(3)$.   Any function $g(r)$ which has real values when $r = 1, 2, 3$

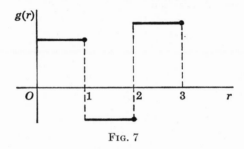

Fig. 7

will represent a vector if it is agreed that these values are the components of the vector.   This function may not be defined for any other values of r, in which case its graph would consist of just three points.

Suppose that $g(r)$, defined when $r = 1, 2$, and 3, is also defined for other values of r to represent this step function (Fig. 7):

$$(1) \qquad g(r) = \begin{cases} g(1) & \text{when } 0 < r \leq 1, \\ g(2) & \text{when } 1 < r \leq 2, \\ g(3) & \text{when } 2 < r \leq 3. \end{cases}$$

Then $[g(1)]^2$ represents the rectangular area under the graph of $[g(r)]^2$ from $r = 0$ to $r = 1$, and similarly for $[g(2)]^2$ and $[g(3)]^2$; thus formula (1), Sec. 23, for the square of the norm can be written

$$\|g\|^2 = \int_0^3 [g(r)]^2 \, dr.$$

If g_1 and g_2 are two such step functions, then

$$(g_1,g_2) = \int_0^3 g_1(r)g_2(r) \, dr.$$

The function $g(r)$ will represent a vector in space of k dimension if it has real values when $r = 1, 2, \ldots, k$, which are considered as the components of the vector.

Now let $g(x)$ be a function defined for all values of x on an interval $a \leqq x \leqq b$. To consider the function as a vector, the components should consist of all the ordinates on its graph, one for each value of x on the interval. It is now impossible to sum with respect to x as we do with respect to the index r. The natural process is to sum by integration.

The *norm* of the function $g(x)$, the length of this generalized vector, is the square root of the sum of squares of its components defined as the nonnegative number

$$(2) \qquad \|g\| = \left\{ \int_a^b [g(x)]^2 \, dx \right\}^{\frac{1}{2}}.$$

The *inner product* of two functions $g_1(x)$ and $g_2(x)$ is the number defined by the equation

$$(3) \qquad (g_1, g_2) = \int_a^b g_1(x)g_2(x) \, dx,$$

in analogy to equation (5), Sec. 24. The condition that the two functions be *orthogonal* is written

$$(g_1, g_2) = 0;$$

that is,

$$(4) \qquad \int_a^b g_1(x)g_2(x) \, dx = 0.$$

As before, definition (2) can be written $\|g\| = (g,g)^{\frac{1}{2}}$.

We have carried our generalization of vectors too far to preserve the original meaning of our geometrical terminology. The norm of a function g has no interpretation as a length associated with g other than the square root of the area under the graph of $[g(x)]^2$. The orthogonality of two functions g_1 and g_2 signifies nothing about perpendicularity, but instead that the product $g_1 g_2$ assumes both negative and positive values on the interval in such a manner that equation (4) is satisfied. The generalized distance between two functions g_1 and g_2,

$$(5) \qquad \|g_1 - g_2\| = \left\{ \int_a^b [g_1(x) - g_2(x)]^2 \, dx \right\}^{\frac{1}{2}},$$

is a measure of mean distance between their graphs.

A set of functions $\{g_n(x)\}$ $(n = 1, 2, \ldots)$ is orthogonal on interval (a,b) if $(g_m, g_n) = 0$ whenever $m \neq n$ $(m = 1, 2, \ldots)$. If none of the functions g_n have zero norms, each function g_n can be normalized by dividing it by the positive constant $\|g_n\|$. The new set $\{\phi_n(x)\}$ so formed, where

$$(6) \qquad \phi_n(x) = \|g_n\|^{-1} g_n(x) \qquad (n = 1, 2, \ldots).,$$

is normed and orthogonal, or *orthonormal* on the interval; that is,

$$(7) \qquad (\phi_m, \phi_n) = \delta_{mn} \qquad (m,n = 1, 2, \ldots),$$

where δ_{mn} is Kronecker's δ (Sec. 24).   Written in full, the characterization (7) of an orthonormal set becomes

$$(8) \quad \int_a^b \phi_m(x) \phi_n(x) \ dx = \begin{cases} 0 & \text{if } m \neq n, \\ 1 & \text{if } m = n. \end{cases} \quad (m, n = 1, 2, \ldots).$$

The interval (a,b) over which the functions and their inner products are defined is called the *fundamental interval*.   Restrictions on the functions, so that the integrals representing their inner products exist, will be noted in the following section.

An example of an orthogonal set of functions was cited in Sec. 17, where we noted that the set

$$(9) \qquad\qquad \left\{ \sin \frac{n\pi x}{c} \right\} \qquad\qquad (n = 1, 2, \ldots)$$

is orthogonal on the interval $(0,c)$.   All functions of this set have the norm $\sqrt{c/2}$ in common, so the orthonormal set is

$$(10) \qquad\qquad \left\{ \sqrt{\frac{2}{c}} \sin \frac{n\pi x}{c} \right\} \quad (0 \leqq x \leqq c, n = 1, 2, \ldots).$$

The set (9) is also orthogonal on the interval $(-c,c)$; in this case the normalizing factor is $c^{-\frac{1}{2}}$.

26. Sectionally Continuous Functions.   Let a function f be continuous at all points of a finite interval $a \leqq x \leqq b$ except possibly for some finite set of points $a, x_1, \ldots, x_n, b$, where $a < x_1 < \cdots < x_n < b$.   Then f is continuous on each of the open intervals $a < x < x_1, x_1 < x < x_2, \ldots, x_n < x < b$; it is not necessarily defined at their end points.   But if in each of those subintervals f has finite limits as x approaches the end points from the interior, then the function f is *sectionally continuous* on the interval (a,b).

Note that the limits $f(a + 0)$, $f(x_1 - 0)$, $f(x_1 + 0)$, . . . , $f(b - 0)$ are required to exist, where such limits from the right and left at a point x_0 are defined as follows: $\epsilon > 0$ and

$$(1) \quad f(x_0 + 0) = \lim_{\epsilon \to 0} f(x_0 + \epsilon), \qquad f(x_0 - 0) = \lim_{\epsilon \to 0} f(x_0 - \epsilon).$$

Continuous functions are special cases of sectionally continuous functions. The step function (1), Sec. 25, illustrated in Fig. 7, is an example of a sectionally continuous function $g(r)$ that is not continuous over the interval $0 \leq r \leq 3$.

The integral of a sectionally continuous function exists. It is the sum of the integrals over the subintervals:

$$(2) \quad \int_a^b f(x)\, dx = \int_a^{x_1} f\, dx + \int_{x_1}^{x_2} f\, dx + \cdots + \int_{x_n}^b f\, dx.$$

The first integral on the right exists because it is the integral of a continuous function over the interval $a \leq x \leq x_1$ if we merely assign the values $f(a + 0)$ and $f(x_1 - 0)$ to $f(a)$ and $f(x_1)$, respectively. Likewise, the other integrals exist as integrals of continuous functions.

If two functions f_1 and f_2 are each sectionally continuous on an interval (a,b), there is a subdivision of the interval such that both functions are continuous on each closed subinterval when the functions are given their limiting values from the interior at the two end points. Hence a linear combination $c_1 f_1 + c_2 f_2$, or the product $f_1 f_2$, has that continuity in each subinterval and is therefore sectionally continuous on (a,b). Consequently the integrals on that interval, of the functions $c_1 f_1 + c_2 f_2$, $f_1 f_2$ and $[f_1(x)]^2$, all exist. Also, one-sided limits of those functions exist as the corresponding combinations of one-sided limits of f_1 and f_2.

Except when otherwise stated in this book, *we shall restrict our attention to functions that are sectionally continuous* on all finite intervals under consideration.

The aggregate of all functions of that class on a given interval (a,b) constitutes the *function space* of sectionally continuous functions. This corresponds to the three-dimensional vector space represented by all functions $g(r)$ which are defined when

$r = 1, 2, 3$.   We have shown above that the integrals used in Sec. 25 to define norms, inner products and generalized distance, exist for all functions in the space of sectionally continuous functions.

Other function spaces can be used in the theory of orthogonal sets of functions.   The subspace consisting of all continuous functions on the interval $a \leq x \leq b$ is simpler, but more restricted than the space of sectionally continuous functions.   The space of all integrable functions f whose products, including squares $[f(x)]^2$, are integrable is used in the general theory of functional analysis.   Our special case involves more elementary concepts in mathematical analysis.

PROBLEMS

1. Show that the set of functions $\{\cos nx\}$ $(n = 0, 1, 2, \ldots)$ is orthogonal on the interval $(0,\pi)$ and that the corresponding orthonormal set consists of the functions $1/\sqrt{\pi}$ and $\sqrt{2/\pi} \cos nx$ $(n = 1, 2, \ldots)$.

2. Show that the functions $\sin x$, $\sin 2x$, $\sin 3x$, $\ldots$, 1, $\cos x$, $\cos 2x$, $\ldots$ constitute an orthogonal set on the interval $(-\pi,\pi)$.   Normalize the set.

3. Show that the functions $f_1(x) = 1$ and $f_2(x) = x$ are orthogonal on the interval $(-1,1)$, and determine the constants A and B so that the function $f_3(x) = 1 + Ax + Bx^2$ is orthogonal to both f_1 and f_2 on that interval.   *Ans. $A = 0$, $B = -3$.*

4. Two continuous functions $f(x)$ and $g_1(x)$ are linearly independent on an interval (a,b); that is, one is not a constant times the other. Determine the linear combination $f + Ag_1$ of those functions which is orthogonal to g_1, thus obtaining the orthogonal pair g_1, g_2 on the interval, where

$$g_2(x) = f(x) - \frac{(f,g_1)}{(g_1,g_1)} g_1(x).$$

Also, note the geometric interpretation of this formula for g_2 in case f, g_1, and g_2 represent actual vectors ($x = 1, 2, 3$).

5. If a function f is continuous on a closed interval, then it is bounded; that is, a number M exists such that $|f(x)| < M$ for all points x on the interval.   Use this fact to state why a sectionally continuous function is bounded for all points on its interval at which the function is defined.

6. For the function $f(x) = 1/\sqrt{x}$, (a) state why f is not sectionally continuous on the interval $(0,1)$: also, (b) show that $\int_0^1 f \, dx$ exists as an improper integral, but that $[f(x)]^2$ is not integrable over the interval $(0,1)$.

7. If $f(x) = 0$ except at a finite number of points on an interval $a \leqq x \leqq b$, state why it is a sectionally continuous function with zero norm, $\|f\| = 0$, on that interval.

8. Given that the integral of a nonnegative continuous function has a positive value if the function has a positive value somewhere on the interval of integration, show that if a function f is sectionally continuous and $\|f\| = 0$ on an interval (a,b), then $f(x) = 0$ at all points x of the interval except possibly for a finite number of points.

27. Generalized Fourier Series. Given an orthonormal set of functions $\{\phi_n(x)\}$ $(n = 1, 2, \ldots)$ on an interval (a,b), it *may* be possible to represent an arbitrarily given function f on that interval by a linear combination of those functions, generalized to an infinite series:

(1) $$f(x) = c_1\phi_1(x) + c_2\phi_2(x) + \cdots + c_n\phi_n(x) + \cdots$$
$$(a < x < b).$$

This corresponds to the representation (2), Sec. 24, of any vector of three-dimensional space in terms of the three vectors in an orthonormal set. We note that the correspondence has its limitations. In Sec. 24 the number of vectors $\phi_n(r)$ $(n = 1, 2, 3)$ in the orthonormal reference set is the same as the number of components of a vector $f(r)$ $(r = 1, 2, 3)$. However, the number of functions $\phi_n(x)$ in the above orthonormal set is countably infinite; that is, there is a one to one correspondence between those functions and the set of all positive integers; still each function $f(x)$ or $\phi_n(x)$ has as many components as there are points x on the interval (a,b), an uncountable infinity of components.

In case the series in equation (1) does converge to $f(x)$ and if, after multiplying all terms in the equation by $\phi_n(x)$, the resulting series is integrable, we can obtain the coefficients c_n as inner products by the process used for vectors. Upon multiplying through by $\phi_n(x)$ and integrating over the interval (a,b) we see that

$$(f,\phi_n) = c_1(\phi_1,\phi_n) + c_2(\phi_2,\phi_n) + \cdots + c_n(\phi_n,\phi_n) + \cdots$$

Since $(\phi_m,\phi_n) = \delta_{mn}$, it follows that

(2) $$(f,\phi_n) = c_n \qquad (n = 1, 2, \ldots)$$

Those numbers c_n are called the *Fourier constants* for f corresponding to the orthonormal set $\{\phi_n\}$; they can be written

$$(3) \qquad c_n = \int_a^b f(x)\phi_n(x)\, dx \qquad (n = 1, 2, \ldots).$$

The series in equation (1) with those coefficients is the *generalized Fourier series* corresponding to the function f, written

$$(4) \qquad f(x) \sim \sum_{n=1}^{\infty} c_n\phi_n(x) = \sum_{n=1}^{\infty} \phi_n(x) \int_a^b f(\xi)\phi_n(\xi)\, d\xi.$$

The correspondence (4) between $f(x)$ and its series will not always be an equality, as we may anticipate by considering the case of vectors in three-dimensional space. In that case if only two vectors $\phi_1(r)$ and $\phi_2(r)$ make up the orthonormal set, any vector not in the plane of those two fails to have a representation of the form $c_1\phi_1(r) + c_2\phi_2(r)$. The reference system here is not complete in the sense that there is a unit vector in three-dimensional space which is perpendicular to both ϕ_1 and ϕ_2.

Likewise in the correspondence (4), in case f is orthogonal to every member ϕ_n of the orthonormal set of functions and $\|f\| \neq 0$, then every term in the series is zero and thus the series does not represent $f(x)$.

An orthonormal set $\{\phi_n(x)\}$ is *complete* in the function space being considered if there is no function in that space, with positive norm, which is orthogonal to each of the functions ϕ_n. We have just noted that the set must necessarily be complete if the correspondence (4) is to be an equality for each function f of the space.

28. Approximation in the Mean.   A linear combination

$$K(r) = \gamma_1\phi_1(r) + \gamma_2\phi_2(r) \qquad (r = 1, 2, 3)$$

of two of the vectors of an orthonormal set $\{\phi_1, \phi_2, \phi_3\}$ is a vector in the plane of ϕ_1 and ϕ_2. To make K the best approximation to a given vector $f(r)$ in the three-dimensional space, in the sense that the distance $d = \|f - K\|$ between the tips of f and K is to be as short as possible, we can see geometrically that K must be the projection of f on the plane of ϕ_1 and ϕ_2 (Fig. 8). Consequently $\gamma_1 = (f, \phi_1) = c_1$, the projection of f on ϕ_1, and $\gamma_2 = (f, \phi_2) = c_2$. Similarly, we see that the coefficients $c_n = (f, \phi_n)$ are those for which a linear combination of any one, two, or three of the vectors ϕ_n best approximates f.

There is a corresponding alternate characterization of the Fourier constants c_n for a function $f(x)$.

Consider functions in the space of sectionally continuous functions on an interval (a,b). Let $\phi_1, \phi_2, \ldots, \phi_m$ denote some

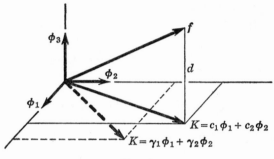

FIG. 8

m functions of an orthonormal set $\{\phi_n\}$ $(n = 1, 2, \ldots)$ on that interval and let K_m be a linear combination of them:

$$(1) \qquad K_m(x) = \gamma_1\phi_1(x) + \gamma_2\phi_2(x) + \cdots + \gamma_m\phi_m(x).$$

We shall determine the constants γ_n so that K_m is the *best approximation in the mean* to a given function f in the sense that the value of the integral E,

$$(2) \qquad E = \int_a^b [f(x) - K_m(x)]^2 \, dx,$$

a measure of the error, is to be as small as possible. This is approximation of $f(x)$ by *least squares*. We note that E is the square of the generalized distance $\|f - K_m\|$ between the functions f and K_m.

Let c_n be the Fourier constants for f, $c_n = (f, \phi_n)$; then

$$E = \int_a^b [f(x) - \gamma_1\phi_1(x) - \gamma_2\phi_2(x) - \cdots - \gamma_m\phi_m(x)]^2 \, dx$$
$$= \int_a^b [f(x)]^2 \, dx + \gamma_1{}^2 + \gamma_2{}^2 + \cdots + \gamma_m{}^2$$
$$- 2\gamma_1 c_1 - 2\gamma_2 c_2 - \cdots - 2\gamma_m c_m.$$

We add and subtract $c_1{}^2, c_2{}^2, \ldots, c_m{}^2$ to complete the squares on the right; thus

$$(3) \quad E = \int_a^b [f(x)]^2 \, dx - c_1^2 - c_2^2 - \cdots - c_m^2 + (\gamma_1 - c_1)^2$$
$$+ (\gamma_2 - c_2)^2 + \cdots + (\gamma_m - c_m)^2.$$

It is clear from equation (2) that $E \geq 0$, so it follows from equation (3) that E has its least value when $\gamma_1 = c_1$, $\gamma_2 = c_2$, . . . , $\gamma_m = c_m$.   The result can be stated as follows.

Theorem 1.   *The Fourier constants of a function f with respect to the functions ϕ_1, ϕ_2, . . . , ϕ_m of an orthonormal set are those coefficients for which a linear combination of those m functions is the best approximation in the mean to $f(x)$, on the fundamental interval.*

Write $\gamma_n = c_n$ in equation (3).   Then since $E \geq 0$,

$$(4) \qquad c_1^2 + c_2^2 + \cdots + c_m^2 \leq \int_a^b [f(x)]^2 \, dx = \|f\|^2;$$

this is known as *Bessel's inequality*.   The number on the right is independent of m, so as m increases the sum on the left is a bounded nondecreasing sequence.   Thus the infinite sequence converges to a limit not greater than $\|f\|^2$; that is, the infinite series of squares of the Fourier constants for each function f of our space converges, and

$$(5) \qquad \sum_{n=1}^{\infty} c_n^2 \leq \int_a^b [f(x)]^2 \, dx.$$

The convergence of this series implies that the general term tends to zero as n becomes infinite.   Hence *the Fourier constants always approach zero as $n \to \infty$* :

$$(6) \qquad \lim_{n \to \infty} c_n = \lim_{n \to \infty} (f, \phi_n) = 0.$$

29. Closed and Complete Sets.   Let the partial sum of the generalized Fourier series corresponding to f be denoted by $S_m(x)$:

$$(1) \qquad S_m(x) = \sum_{n=1}^{m} c_n \phi_n(x).$$

This is the linear combination $K_m(x)$ in Sec. 28 when $\gamma_n = c_n$.

The sum S_m is said to *converge in the mean* to the function f, over the fundamental interval, if

$$(2) \qquad \lim_{m \to \infty} \int_a^b [f(x) - S_m(x)]^2 \, dx = 0.$$

Condition (2) is also written

$$\operatorname*{l.i.m.}_{m \to \infty} S_m(x) = f(x),$$

where the abbreviation l.i.m. stands for *limit in the mean*.

If condition (2) is satisfied by each function f in our function space, we call the orthonormal set $\{\phi_n(x)\}$ *closed in the sense of mean convergence*. Each function f can be approximated arbitrarily closely, in the mean, by some linear combination of functions ϕ_n of such a closed set.[1]

By expanding the integrand in equation (2) and keeping the definition of c_n in mind, we can write that equation in the form

$$(3) \qquad \lim_{m \to \infty} \left\{ \int_a^b [f(x)]^2 \, dx - 2 \sum_{n=1}^m c_n^2 + \sum_{n=1}^m c_n^2 \right\} = 0.$$

Hence for a closed set $\{\phi_n\}$ it is true that

$$(4) \qquad \sum_{n=1}^\infty c_n^2 = \int_a^b [f(x)]^2 \, dx.$$

This is known as *Parseval's equation*. When written in the form

$$(5) \qquad \sum_{n=1}^\infty (f, \phi_n)^2 = \|f\|^2$$

it identifies the sum of the squares of the components of f with respect to the generalized reference vectors ϕ_n, with the square of the norm of f.

Conversely, if each function f of the space satisfies Parseval's equation, the set $\{\phi_n\}$ is closed in the sense of mean convergence. This is true because equations (3) and (4) are merely alternate forms of equation (2).

---

[1] In the mathematical literature, the terms closed and complete are sometimes applied to sets which we have called complete and closed, respectively.

Suppose that a function θ in the space, where $\|\theta\| \neq 0$, is orthogonal to each function ϕ_n of the closed orthonormal set. Then equation (5), with f replaced by θ, gives the contradiction $\|\theta\| = 0$; thus the set is complete (Sec. 27):

Theorem 2. *If an orthonormal set* $\{\phi_n(x)\}$ *is closed, it is complete.*

This bare introduction to the theory of general orthogonal sets, based on convergence in the mean, will not be continued here. *Convergence in the mean does not ensure pointwise convergence;* that is, the statement (2) is not the same as the statement

$$(6) \qquad\qquad \lim_{m \to \infty} S_m(x) = f(x)$$

for each point x of the interval (a,b), even if some points may be excepted.[1] We shall be concerned with pointwise convergence.

An orthonormal set is *closed in the sense of pointwise convergence*, let us say, if its generalized Fourier series for each function f of our function space converges pointwise to $f(x)$, except possibly at a finite number of points of the fundamental interval. At the end of Sec. 27 we indicated why a set that is closed in that sense must be complete. Thus Theorem 2 is true for such closed sets, in the space of sectionally continuous functions.

The direct application of Theorem 2 is limited, however, to denying that a set is closed in case there is a function with positive norm which is orthogonal to each member of the set. It is our representation theorems which will show that certain sets *are* closed, and hence complete, in specified function spaces.

PROBLEMS

1. Show that the set $\{\sqrt{2/c} \cos (n\pi x/c)\}$ $(n = 1, 2, \ldots)$ is orthonormal on the interval $0 \leqq x \leqq c$, but not complete even in the space of continuous functions with continuous derivatives without the inclusion of some constant function corresponding to the case $n = 0$. (We shall see later that the larger set is complete in the space.)

2. Show why the set $\{\sin n\pi x\}$, $n = 1, 2, \ldots$ is orthogonal on the interval $-1 \leqq x \leqq 1$, but not complete even in the space of all continuous functions on that interval.

---

[1] A simple example of a sequence of functions which converges in the mean to zero, but which *diverges at each point* of the interval, is given by P. Franklin, "Treatise on Advanced Calculus," p. 408, 1940.

3. If a function f has a representation

$$f(x) = \sum_{n=1}^{\infty} A_n g_n(x)$$

on a fundamental interval (a,b) on which the set $\{g_n\}$ is orthogonal but not normalized, $0 < \|g_n\| \neq 1$, use inner products to show formally that

$$A_n = \frac{(f,g_n)}{\|g_n\|^2}.$$

Also note that the above series with these coefficients is the generalized Fourier series (4), Sec. 27, where $\phi_n = g_n/\|g_n\|$.

4. Verify that

$$\tfrac{1}{2}\int_a^b \int_a^b [f(x)g(y) - g(x)f(y)]^2 dx\, dy = \|f\|^2\|g\|^2 - (f,g)^2,$$

assuming that f and g are sectionally continuous. Thus establish the *Schwarz inequality*

$$(f,g)^2 \leqq (f,f)(g,g).$$

Also, note the corresponding inequality for vectors in three-dimensional space.

5. Use the Schwarz inequality (Problem 4) to prove that, if f and g are sectionally continuous functions on a fundamental interval (a,b) and if either has zero norm, say $\|f\| = 0$, then $(f,g) = 0$.

6. Consider continuous functions on a fundamental interval $a \leqq x \leqq b$. Then the norm of a function vanishes if and only if the function vanishes at every point of the interval. If two functions f and g have the same set of Fourier constants with respect to a *complete* orthonormal set $\{\phi_n\}$, $c_n = (f,\phi_n) = (g,\phi_n)$, $n = 1, 2, \ldots$, show why the functions must be identical; that is, $f(x) - g(x) = 0$, $a \leqq x \leqq b$. This is a uniqueness property for functions with components c_n.

7. Consider continuous functions on a fundamental interval $a \leqq x \leqq b$. Let $\{\phi_n\}$ be a complete orthonormal set. If the corresponding generalized Fourier series for a function f converges uniformly over the interval, prove that its sum $S(x)$ is identical to $f(x)$. (See Problem 6). Also, note why the conclusion holds if the condition of completeness is replaced by the condition that the set is closed in the sense of mean convergence.

8. Show that the sequence of functions f_n, $n = 1, 2, \ldots$, where $f_n(x) = 0$ when $0 \leqq x \leqq 1/n$, $f_n(x) = \sqrt{n}$ when $1/n < x < 2/n$, and $f_n(x) = 0$ when $2/n \leqq x \leqq 1$, converges to zero at each point x of the interval $0 \leqq x \leqq 1$; but that the sequence does not converge in the mean to zero, over that interval.

30. Complex-valued Functions.   We shall have a few occasions to use complex-valued functions of a real variable, functions of the type

$$(1) \qquad\qquad w(t) = u(t) + iv(t),$$

where u and v are real-valued functions of a real variable t.

The derivative of w is defined in this natural way:

$$(2) \qquad\qquad w'(t) = u'(t) + iv'(t),$$

provided u and v are differentiable.   Similarly,

$$\int_a^b w(t) \, dt = \int_a^b u(t) \, dt + i \int_a^b v(t) \, dt.$$

The operation $\bar{w} = u - iv$ of taking the complex conjugate of w commutes with the operations of differentiation and integration; that is, for differentiation, $(\bar{w})' = \overline{w'}$.   From equation (2) we can show that the elementary rules of differentiation for real-valued functions, such as the formula for the derivative of a product, apply to these complex-valued functions.

The exponential function exp z, or e^z, is defined by

$$(3) \qquad\qquad \exp z = e^x(\cos y + i \sin y),$$

where $z = x + iy$, x and y real, and the angle y means y radians. From this definition it can be seen that e^z satisfies the usual laws of exponents.   Also, if x and y are functions of a real variable t, then z and exp z are functions of type (1) and, with the aid of equation (2), we can show that

$$(4) \qquad\qquad \frac{d}{dt}\,(e^z) = e^z \frac{dz}{dt}.$$

Thus, for example, the function $w = \exp(ict)$ satisfies the differential equation $w''(t) + c^2 w(t) = 0$ when c is a complex number.

From the definitions

$$(5) \qquad \sin z = \frac{1}{2i}\,(e^{iz} - e^{-iz}), \qquad \cos z = \tfrac{1}{2}(e^{iz} + e^{-iz})$$

and definition (3), we find that

$$(6) \quad |\sin z|^2 = \sin^2 x + \sinh^2 y, \qquad |\cos z|^2 = \cos^2 x + \sinh^2 y$$

and, when $z = z(t)$, that

(7) $\qquad \dfrac{d}{dt} \sin z = \cos z \dfrac{dz}{dt}, \qquad \dfrac{d}{dt} \cos z = -\sin z \dfrac{dz}{dt}.$

Note that $2i \sin t = e^{it} - e^{-it}$, which agrees with formula (3). The hyperbolic functions

(8) $\qquad \sinh z = \tfrac{1}{2}(e^z - e^{-z}), \qquad \cosh z = \tfrac{1}{2}(e^z + e^{-z})$

are clearly related to the trigonometric functions as follows:

(9) $\qquad \sin iz = i \sinh z, \qquad \cos iz = \cosh z.$

For any angle θ we can write

$$2(\cos \theta + \cos 2\theta + \cdots + \cos N\theta) = \sum_{n=1}^{N} e^{in\theta} + \sum_{n=1}^{N} e^{-in\theta}.$$

The sum of the two finite geometric series on the right is

$$\frac{e^{i\theta}(1 - e^{iN\theta})}{1 - e^{i\theta}} + \frac{e^{-i\theta}(1 - e^{-iN\theta})}{1 - e^{-i\theta}}$$

if $e^{i\theta} \neq 1$. This can be written

$$\frac{-\exp\left(\tfrac{1}{2}i\theta\right) + \exp\left[i\theta(N + \tfrac{1}{2})\right] + \exp\left(-\tfrac{1}{2}i\theta\right) - \exp\left[-i\theta(N + \tfrac{1}{2})\right]}{\exp\left(\tfrac{1}{2}i\theta\right) - \exp\left(-\tfrac{1}{2}i\theta\right)}.$$

Thus we obtain *Lagrange's trigonometric identity*

(10) $\qquad 2 \displaystyle\sum_{n=1}^{N} \cos n\theta = -1 + \dfrac{\sin\left[(N + \tfrac{1}{2})\theta\right]}{\sin \tfrac{1}{2}\theta} \qquad \left(\sin \dfrac{\theta}{2} \neq 0\right).$

This identity will be used in the theory of Fourier series.

31. Other Types of Orthogonality. Extensions of the concept of orthogonal sets of functions should be noted.

a. A set $\{g_n(x)\}$ is orthogonal on an interval (a,b) with respect to a *weight function* $p(x)$, where $p(x) \geqq 0$, if

(1) $\qquad \displaystyle\int_a^b p(x)g_m(x)g_n(x)\,dx = 0 \qquad$ when $m \neq n$.

The integral here represents the inner product (g_m, g_n) with respect to the weight function, corresponding to the inner product of vectors when weight numbers are used (Sec. 24). The set is normalized by dividing g_n by $\|g_n\|$, where

$$\|g_n\|^2 = (g_n, g_n) = \int_a^b p(x)[g_n(x)]^2 \, dx.$$

This type of orthogonality is reduced to the ordinary type by using the products $\sqrt{p(x)}g_n(x)$ as the functions of the set.

Weight functions will arise in orthogonal sets of Bessel functions and in some sets generated by Sturm-Liouville problems. Another example is the set of *Tchebysheff polynomials*,

$$(2) \qquad\qquad T_n(x) = \cos{(n \arccos x)} \quad (n = 0, 1, 2, \ldots),$$

which is orthogonal on the interval $(-1,1)$ with respect to the weight function

$$(3) \qquad\qquad p(x) = (1 - x^2)^{-\frac{1}{2}}.$$

b. A set $\{w_n\}$ of complex-valued functions (Sec. 30) of a real variable x is orthogonal in the *hermitian sense*, on an interval (a,b), if

$$(4) \qquad\qquad \int_a^b w_m(x)\overline{w_n(x)} \, dx = 0 \qquad\qquad (m \neq n).$$

The integral here is the hermitian inner product (w_m, w_n). The square of the norm of w_n is real and nonnegative since

$$(5) \qquad \|w_n\|^2 = \int_a^b w_n \bar{w}_n \, dx = \int_a^b (u_n^2 + v_n^2) \, dx$$

if $w_n = u_n + iv_n$, where u_n and v_n are real-valued functions of x.

Certain complex-valued exponential functions furnish the most prominent examples of such sets. The functions

$$(6) \qquad\qquad e^{inx} = \cos nx + i \sin nx \quad (n = 0, \pm 1, \pm 2, \ldots)$$

constitute a set with hermitian orthogonality on the interval $(-\pi,\pi)$. The proof is left to the problems.

c. For sets $\{g_n(x,y)\}$ of functions of two independent variables the fundamental interval is replaced by a region in the xy plane, and the integrations are made over that region. Similar extensions apply to functions of three or more variables.

PROBLEMS

1. Verify that solutions of the differential equation

$$w''(t) + \lambda w(t) = 0,$$

in which λ is any complex constant, may be written in each of the forms

$$w = C_1 \exp (it \sqrt{\lambda}) + C_2 \exp (-it \sqrt{\lambda}) = C_3 \sin t \sqrt{\lambda} + C_4 \cos t \sqrt{\lambda}$$
$$= \tfrac{1}{2} C_5 \{\exp [i(t \sqrt{\lambda} + C_6)] + \exp [-i(t \sqrt{\lambda} + C_6)]\}$$
$$= C_5 \cos (t \sqrt{\lambda} + C_6),$$

where $C_1, C_2, \ldots, C_6$ are arbitrary complex constants.

2. Use equations (6), Sec. 30, to show (a) that the zeros of the functions $\sin z$ and $\cos z$ are all real; (b) that $|\sin z|$ and $|\cos z|$ are unbounded as $|y| \to \infty$.

3. Establish the hermitian orthogonality of the exponential functions (6), Sec. 31, on the interval $(-\pi, \pi)$. Also show that the norm of each function is $\sqrt{2\pi}$.

4. Show that the set $\{(b - a)^{-\frac{1}{2}} \exp [2n\pi i x/(b - a)]\}, n = 0, \pm 1, \pm 2,$ . . . is orthonormal in the hermitian sense on the interval (a, b).

5. Substitute $\theta = \arccos x$ to establish the orthogonality of the Tchebysheff polynomials (2) with weight function (3), Sec. 31.

6. Write $x = \cos \theta$ in formula (2), Sec. 31, for the functions $T_n(x)$. In De Moivre's formula

$$(\cos \theta + i \sin \theta)^n = \cos n\theta + i \sin n\theta$$

use the binomial expansion and equate real parts to show that $\cos n\theta$ is a polynomial of degree n in $\cos \theta$, and hence that $T_n(x)$ is actually a polynomial of degree n in x.

32. Sturm-Liouville Problems.

In Sec. 16 a method of separation of variables, and superposition of functions of type $X(x) T(t)$ that satisfy the homogeneous conditions, was used to write a formal solution of a boundary value problem in the displacements $y(x, t)$ of a stretched string. The process required the function X to be a solution of the homogeneous problem

$$(1) \qquad X''(x) + \lambda X(x) = 0, \qquad X(0) = 0, \qquad X(c) = 0.$$

For a discrete set of values of the parameter $\lambda = n^2 \pi^2/c^2$, $n = 1, 2, \ldots$ we found that problem (1) has nontrivial solutions $X = \sin (n\pi x/c)$ and that those functions are orthogonal on the interval $(0, c)$.

When applied to more general boundary value problems in partial differential equations, the process frequently leads to a homogeneous differential equation of type

(2) $X''(x) + R(x)X'(x) + [Q(x) + \lambda P(x)]X(x) = 0$

involving a parameter λ in the manner indicated and to a pair of homogeneous boundary conditions of type

(3) $a_1 X(a) + a_2 X'(a) = 0, \qquad b_1 X(b) + b_2 X'(b) = 0.$

The functions P, Q, and R and the constants a_1, a_2, b_1, and b_2 are prescribed by the problem in partial differential equations; λ and $X(x)$ are to be determined.

Other types of homogeneous two-point boundary value problems involving a parameter may arise instead of the problem consisting of equations (2) and (3).   But the problem (2) and (3), the *Sturm-Liouville problem*, is of fundamental importance.[1]

When its terms are multiplied by a function

$$r(x) = \exp [\int R(x) \, dx],$$

an integrating factor for $X'' + RX'$, the Sturm-Liouville equation (2) takes the form

(4) $$\frac{d}{dx}\left[r(x)\frac{dX}{dx} \right] + [q(x) + \lambda p(x)]X = 0.$$

Under rather general conditions on the functions p, q, and r it can be shown that there is always a countable infinity of values $\lambda_1, \lambda_2, \ldots$ of the parameter λ for each of which the Sturm-Liouville problem (3) and (4) has a solution that is not identically zero.   The numbers λ_n are the *eigenvalues*, or *characteristic numbers*, of the problem, and the corresponding solutions $X_n(x)$ are the *eigenfunctions*, or *characteristic functions*.   Note that CX_n is also an eigenfunction, where C is any constant other than zero.   In the special case (1), $\lambda_n = n^2\pi^2/c^2$ and $X_n = \sin (n\pi x/c)$.

The orthogonality of the eigenfunctions with weight function *p*, on the interval (a,b), is established in the following section. With respect to the set of normalized eigenfunctions

$$\phi_n(x) = X_n(x)\|X_n\|^{-1} \quad \text{where } \|X_n\|^2 = \int_a^b pX_n{}^2 \, dx$$

[1] Papers by J. C. F. Sturm and J. Liouville giving the first extensive development of the theory of this problem appeared in the first three volumes of *Journal de mathématique*, 1836–1838.

the generalized Fourier series for a function f is

$$(5) \qquad \sum_{n=1}^{\infty} c_n \phi_n(x) \qquad \text{where } c_n = \int_a^b pf\phi_n \, dx.$$

For prominent special cases and singular cases of the Sturm-Liouville problem we shall establish the convergence of this series to $f(x)$ on the interval (a,b).

The representation theory for the general case is not presented in this book.[1] For the general case proofs that series (5) converges to $f(x)$ usually employ either the theory of residues of functions of a complex variable or a comparison of the series with an ordinary Fourier series that represents f. Proofs are complicated by the fact that explicit solutions of the Sturm-Liouville equation with arbitrary coefficients cannot be written. Properties of solutions are found, by interesting and useful devices, from the differential equation itself.

In case $r(a) = r(b)$ the statements made above hold true when boundary conditions (3) are replaced by the *periodic boundary conditions*

$$(6) \qquad X(a) = X(b), \qquad X'(a) = X'(b).$$

Such conditions commonly arise when x represents a coordinate such as the angle ϕ in cylindrical coordinates, on an interval $(0,2\pi)$.

The *adjoint* of a linear differential operator M of second order, where

$$(7) \qquad M[X(x)] = A(x)X''(x) + B(x)X'(x) + C(x)X(x),$$

is the operator M^* such that

$$(8) \qquad M^*[X(x)] = [A(x)X(x)]'' - [B(x)X(x)]' + C(x)X(x).$$

The operator L, where

$$(9) \qquad L[X(x)] = [r(x)X'(x)]' + q(x)X(x),$$

[1] The general case is treated in chap. 9 of "Operational Mathematics," by R. V. Churchill. For other treatments of Sturm-Liouville theory, some quite extensive, see the books by Ince, Coddington and Levinson, and Titchmarsh ("Eigenfunction Expansions") listed in our Bibliography. Other aspects of the subject are introduced in the books by Collatz and Friedman.

is *self-adjoint;* that is, L and L^* are the same since

$$rX'' + r'X' + qX = (rX)'' - (r'X)' + qX.$$

Form (4) of the Sturm-Liouville equation turns out to be especially useful because it is written in terms of the self-adjoint operator L; it has the form

$$(10) \qquad L[X(x)] = -\lambda p(x)X(x).$$

Some properties of L will be noted explicitly in the problems.

33. Orthogonality of the Eigenfunctions. A few results in the general theory can be established here. They will be useful in the chapters to follow. In special cases the eigenvalues and eigenfunctions will be found, so their existence will not be in doubt.

We assume, unless otherwise stated, that the coefficients in the Sturm-Liouville problem

$$(1) \qquad (rX')' + (q + \lambda p)X = 0,$$

$$(2) \qquad a_1X(a) + a_2X'(a) = 0, \qquad b_1X(b) + b_2X'(b) = 0$$

satisfy these conditions: *p, q, r,* and *r'* are *real-valued continuous functions of* x $(a \leqq x \leqq b)$; *also* $p(x) > 0$ *and* $r(x) > 0$ *when* $a < x < b$, *and the constants* a_1, a_2, b_1, *and* b_2 *are real and independent of* λ. Eigenfunctions X_n are to satisfy the regularity conditions usually required of solutions of differential equations of second order (Sec. 19), namely, that X_n *and* X_n' *be continuous on the interval* $a \leqq x \leqq b$.

Theorem 3. *Let* λ_m *and* λ_n *be any two distinct eigenvalues of the Sturm-Liouville problem* (1) *and* (2), *with corresponding eigenfunctions* X_m *and* X_n. *Then* X_m *and* X_n *are orthogonal with weight function* p, *on the interval* (a,b). *The orthogonality holds also in the following cases:* (a) *when* $r(a) = 0$, *in which case the first of boundary conditions* (2) *can be dropped from the problem;* (b) *when* $r(b) = 0$, *in which case the second of those conditions can be dropped;* (c) *if* $r(a) = r(b)$ *and conditions* (2) *are replaced by the periodic boundary conditions* (6), *Sec. 32.*

To prove this theorem we first note that

$$(rX_m')' + qX_m = -\lambda_m pX_m, \qquad (rX_n')' + qX_n = -\lambda_n pX_n$$

since each eigenfunction satisfies equation (1) when λ has the corresponding eigenvalue. We multiply the members of these two equations by X_n and X_m, respectively, and subtract to get the relation

$$(3) \qquad (\lambda_m - \lambda_n)pX_mX_n = (rX_n')'X_m - (rX_m')'X_n$$
$$= \frac{d}{dx}[(rX_n')X_m - (rX_m')X_n].$$

The final reduction here to an exact derivative was possible because of the self-adjoint operator involved.

Our continuity conditions now permit us to write

$$(4) \qquad (\lambda_m - \lambda_n)\int_a^b pX_mX_n\,dx = [(X_mX_n' - X_nX_m')r]_{x=a}^{x=b}.$$

The last expression in parentheses is the value of the determinant

$$(5) \qquad \Delta(x) = \begin{vmatrix} X_m(x) & X_m'(x) \\ X_n(x) & X_n'(x) \end{vmatrix};$$

therefore

$$(6) \qquad (\lambda_m - \lambda_n)\int_a^b pX_mX_n\,dx = r(b)\,\Delta(b) - r(a)\Delta(a).$$

The first of boundary conditions (2) requires that

$$a_1X_m(a) + a_2X_m'(a) = 0, \qquad a_1X_n(a) + a_2X_n'(a) = 0,$$

and for these simultaneous equations in a_1 and a_2 to be satisfied by numbers a_1 and a_2, not both zero, it is necessary that the determinant $\Delta(a)$ be zero. Similarly, from the second boundary condition, where b_1 and b_2 are not both zero, we see that $\Delta(b) = 0$. Then according to equation (6),

$$(7) \qquad (\lambda_m - \lambda_n)\int_a^b pX_mX_n\,dx = 0$$

and, since $\lambda_m \neq \lambda_n$, the orthogonality property follows:

$$(8) \qquad \int_a^b p(x)X_m(x)X_n(x)\,dx = 0 \qquad (\lambda_m \neq \lambda_n).$$

In case $r(a) = 0$, condition (8) follows from equation (6) when $\Delta(a) \neq 0$; that is, when $a_1 = a_2 = 0$, in which case the first of boundary conditions (2) disappears. Similarly, if $r(b) = 0$, the second of those conditions is not needed.

In case $r(a) = r(b)$ and the periodic conditions $X(a) = X(b)$,

$X'(a) = X'(b)$ are used in place of conditions (2), then

$$r(b) \, \Delta(b) = r(a) \, \Delta(a)$$

and again equation (8) follows. This completes the proof of Theorem 3.

Suppose that X is an eigenfunction corresponding to an eigenvalue $\lambda = \alpha + i\beta$, where α and β are real numbers. Then $X(x)$, which may be complex-valued, satisfies equations (1) and (2). Taking complex conjugates of all terms in those equations and recalling which coefficients are real-valued, we see that (Sec. 30)

$$(r\bar{X}')' + (q + \bar{\lambda}p)\bar{X} = 0,$$
$$a_1\bar{X}(a) + a_2\bar{X}'(a) = 0, \qquad b_1\bar{X}(b) + b_2\bar{X}'(b) = 0.$$

Thus $\bar{X}$ is an eigenfunction corresponding to the eigenvalue $\bar{\lambda}$ and, according to equation (7),

$$(\lambda - \bar{\lambda}) \int_a^b p(x)X(x)\overline{X(x)} \, dx = 0.$$

But $p(x) > 0$ when $a < x < b$ and $X\bar{X} = |X|^2$, so the integral here has a positive value. Since $\lambda - \bar{\lambda} = 2i\beta$, it follows that $\beta = 0$; that is, λ is real.

The argument also applies to cases (a) to (c) in Theorem 3.

Theorem 4. *For the Sturm-Liouville problem, or its modifications (a), (b), or (c) cited in Theorem 3, each eigenvalue is real.*

34. Uniqueness of Eigenfunctions. Let X and Y denote two eigenfunctions of the Sturm-Liouville problem (1) and (2), Sec. 33, corresponding to the same real eigenvalue λ. Suppose that r does not vanish at one end point of the interval, say $r(a) > 0$.

The linear combination

(1) $$W(x) = Y'(a)X(x) - X'(a)Y(x)$$

satisfies the homogeneous differential equation

(2) $$(rW')' + (q + \lambda p)W = 0;$$

also $W'(a) = 0$. Since X and Y satisfy the equations

(3) $$a_1X(a) + a_2X'(a) = 0, \qquad a_1Y(a) + a_2Y'(a) = 0,$$

where a_1 and a_2 are not both zero and $W(a)$ is the determinant of that pair of equations in a_1 and a_2, then $W(a) = 0$. According to the uniqueness theorem for solutions of linear differential equations (Sec. 19), $W(x) = 0$ is the only solution of equation (2) for which $W(a) = W'(a) = 0$. Thus

$$(4) \qquad Y'(a)X(x) - X'(a)Y(x) = 0.$$

Unless $Y'(a) = X'(a) = 0$, equation (4) states that the functions X and Y are linearly dependent, that one is a constant times the other. Recall that zero is not an eigenfunction. If $Y'(a) = X'(a) = 0$, then $a_1 = 0$ in equations (3). In that case we use the linear combination

$$(5) \qquad Z(x) = Y(a)X(x) - X(a)Y(x)$$

to prove in like manner that $Z(x) = 0$ and hence that X and Y are linearly dependent.

Suppose that $X = u + iv$ is a complex-valued eigenfunction corresponding to a real eigenvalue λ. When we substitute $u + iv$ for X in the Sturm-Liouville problem and separate real and imaginary parts, we see that u and v are each eigenfunctions corresponding to λ. Hence $v = ku$ and $X = (1 + ik)u$, where k is a constant. That is, X is real except possibly for an imaginary constant factor.

We collect our results as follows.

Theorem 5. *Under the additional condition that either $r(a) > 0$ or $r(b) > 0$, the Sturm-Liouville problem (1) and (2), Sec. 33, cannot have two linearly independent eigenfunctions that correspond to the same eigenvalue; also, each eigenfunction can be made real-valued by multiplying it by an appropriate nonzero constant.*

That Theorem 5 does not always apply when conditions (2), Sec. 33, are replaced by the periodic boundary conditions is shown by this important special case:

$$(6) \qquad X'' + \lambda X = 0, \qquad X(-\pi) = X(\pi), \qquad X'(-\pi) = X'(\pi).$$

When $\lambda \neq 0$, the solution of the differential equation here is

$$X(x) = C_1 \sin x \sqrt{\lambda} + C_2 \cos x \sqrt{\lambda}.$$

If it is to satisfy both boundary conditions, we find that

$$(7) \qquad C_1 \sin \pi \sqrt{\lambda} = 0 \qquad \text{and} \qquad C_2 \sin \pi \sqrt{\lambda} = 0.$$

Since C_1 and C_2 must not both vanish if X is to be an eigenfunction, it follows that $\lambda = n^2$ $(n = 1, 2, \ldots)$, while the constants C_1 and C_2 are arbitrary.

In particular, for any constant B_n the two functions

$$(8) \qquad X_n(x) = \sin nx, \qquad Y_n(x) = B_n \sin nx + \cos nx$$

are linearly independent eigenfunctions corresponding to the same eigenvalue $\lambda = n^2$. Other linear combinations of $\sin nx$ and $\cos nx$ can be used to form such pairs.

When two functions are linearly independent, there is always a linear combination of the two which is orthogonal to one of the functions (Problem 4, Sec. 26). If $B_n = 0$ in equations (8), Y_n is orthogonal to X_n on the interval $(-\pi,\pi)$.

A constant is the only eigenfunction of problem (6) corresponding to the eigenvalue $\lambda = 0$. According to Theorem 3, eigenfunctions corresponding to distinct eigenvalues are orthogonal. Here $p(x) = 1$. Consequently the set

$$(9) \qquad \{\sin x, \sin 2x, \sin 3x, \ldots, 1, \cos x, \cos 2x, \ldots\},$$

a set of eigenfunctions of problem (6), is orthogonal on the interval $(-\pi,\pi)$. Note that this includes the orthogonality of $\sin nx$ and $\cos nx$, each of which corresponds to the same eigenvalue $\lambda = n^2$.

PROBLEMS

Find all eigenvalues and eigenfunctions of the Sturm-Liouville problems presented in Problems 1 to 6. Also note the interval and weight function p in the orthogonality relation ensured by Theorem 3.

1. $X'' + \lambda X = 0$; $X(0) = 0$, $X'(\tfrac{1}{2}\pi) = 0$.

$\qquad\qquad$ *Ans.* $X_n = \sin[(2n - 1)x]$ $(n = 1, 2, \ldots)$.

2. $X'' + \lambda X = 0$; $X'(0) = 0$, $X'(c) = 0$.

$\qquad\qquad$ *Ans.* $X_n = \cos(n\pi x/c)$ $(n = 0, 1, 2, \ldots)$.

3. $X'' + \lambda X = 0$; $X'(-\pi) = 0$, $X'(\pi) = 0$.

$\qquad\qquad$ *Ans.* $X_n = \cos \dfrac{n(\pi + x)}{2}$ $(n = 0, 1, 2, \ldots)$.

4. $(x^3 X')' + \lambda x X = 0$; $X(1) = 0$, $X(e) = 0$. Note that the differential equation reduces to one of Cauchy type after the indicated differentiation is performed.

$\qquad\qquad$ *Ans.* $X_n = x^{-1} \sin(n\pi \log x)$, $n = 1, 2, \ldots$; $p(x) = x$.

5. $X'' + \lambda X = 0$; $X(0) = 0$, $X(1) - X'(1) = 0$. *Suggestions:* Substitute $\lambda = 0$ to show that $X_0(x) = x$ is an eigenfunction $(\lambda_0 = 0)$.

Write $\lambda = -\alpha^2$, α real, and use graphs of $y = \alpha$ and $y = \tanh \alpha$ to see that no negative eigenvalues exist. Write $\lambda = \alpha^2$ and use graphs of $y = \alpha$ and $y = \tan \alpha$ to see that the positive eigenvalues are $\lambda_n = \alpha_n{}^2$ $(n = 1, 2, \ldots)$, where α_n are the positive roots of the equation $\tan \alpha = \alpha$.

Ans. $X_0 = x$, $X_n = \sin \alpha_n x$ $(n = 1, 2, \ldots)$.

6. $X'' + \lambda X = 0$; $X(0) = 0$, $hX(1) + X'(1) = 0$ where h is constant and $h > -1$. (Compare Problem 5.) Show that the eigenvalues consist only of the numbers $\lambda_n = \alpha_n{}^2$, where α_n are the positive roots of the equation $h \tan \alpha = -\alpha$, and that $X_n = \sin \alpha_n x$ $(n = 1, 2, \ldots)$.

7. When $h < -1$ in Problem 6, show that $\lambda_0 = -\alpha_0{}^2$ is an eigenvalue with eigenfunction $X_0 = \sinh \alpha_0 x$, where $\alpha = \alpha_0$ is the positive root of the equation $\tanh \alpha = -\alpha/h$; also show that $\lambda_n = \alpha_n{}^2$ and $X_n = \sin \alpha_n x$ $(n = 1, 2, \ldots)$, where $\tan \alpha_n = -\alpha_n/h$.

8. In Sec. 34, use the function Z defined by equation (5) to prove that X and Y are linearly dependent, in the case $a_1 = 0$.

9. Give the steps that lead to equations (7), Sec. 34.

10. For the eigenvalue problem (6) in Sec. 34, if α (radians) is a constant angle such that $\sin \alpha \neq 0$, note that the two linearly independent functions $\sin nx$ and $\sin (nx + \alpha)$ are eigenfunctions corresponding to the same eigenvalue $\lambda = n^2$ $(n = 1, 2, \ldots)$. Show that their linear combination $\cos (nx + \alpha)$ is orthogonal to $\sin (nx + \alpha)$ (see Problem 4, Sec. 26) and hence that the functions

$$\sin (x + \alpha), \sin (2x + \alpha), \ldots, 1, \cos (x + \alpha), \cos (2x + \alpha), \ldots$$

constitute an orthogonal set of eigenfunctions on the interval $(-\pi, \pi)$.

11. For the eigenvalue problem

$$X'' + \lambda X = 0; \qquad X(0) = X(2c), \qquad X'(0) = X'(2c),$$

obtain this set of eigenfunctions orthogonal on the interval $(0, 2c)$:

$$\left\{ \sin \frac{m\pi x}{c}, 1, \cos \frac{n\pi x}{c} \right\} \qquad (m, n = 1, 2, \ldots).$$

12. Prove that the self-adjoint operator L defined by equation (9), Sec. 32, satisfies *Lagrange's identity for that operator*

$$XL[Y] - YL[X] = \frac{d}{dx} [(XY' - X'Y)r]$$

for each pair of functions X, Y, assuming that all derivatives exist.

13. If X and Y in Problem 12 satisfy the boundary conditions in the Sturm-Liouville problem or the periodic boundary conditions, show that

$$(X, L[Y]) = (Y, L[X]),$$

where these inner products on the interval (a,b) are without weight functions.

14. If N is the operator d^4/dx^4, show that

$$XN[Y] - YN[X] = \frac{d}{dx}(XY''' - YX''' - X'Y'' + Y'X'').$$

Hence show that if X_1 and X_2 are eigenfunctions of the eigenvalue problem of fourth order

$$N[X] = -\lambda X; \qquad X(0) = X''(0) = 0, \qquad X(c) = X''(c) = 0,$$

corresponding to distinct eigenvalues λ_1 and λ_2, then X_1 is orthogonal to X_2 on the interval $(0,c)$.

CHAPTER 4

FOURIER SERIES

35. The Basic Series. The trigonometric series

$$(1) \qquad \tfrac{1}{2}a_0 + \sum_{n=1}^{\infty} (a_n \cos nx + b_n \sin nx)$$

is a *Fourier series* if its coefficients are given by the formulas

$$(2) \qquad \begin{aligned} a_n &= \frac{1}{\pi} \int_{-\pi}^{\pi} f(x) \cos nx \, dx \quad (n = 0, 1, 2, \ldots), \\ b_n &= \frac{1}{\pi} \int_{-\pi}^{\pi} f(x) \sin nx \, dx \qquad (n = 1, 2, \ldots), \end{aligned}$$

where f is some function defined on the interval $(-\pi,\pi)$.

In Sec. 34 we noted that the functions 1, $\cos x$, $\cos 2x$, $\cos 3x$, $\ldots$, $\sin x$, $\sin 2x$, $\ldots$ constitute a set of eigenfunctions, orthogonal on the interval $(-\pi,\pi)$, for the eigenvalue problem

$$(3) \qquad \begin{aligned} X''(x) + \lambda X(x) &= 0; \\ X(-\pi) = X(\pi), \qquad X'(-\pi) &= X'(\pi). \end{aligned}$$

The normalized orthogonal set is

$$(4) \qquad \left\{ \frac{1}{\sqrt{2\pi}}, \frac{\cos mx}{\sqrt{\pi}}, \frac{\sin nx}{\sqrt{\pi}} \right\} \quad (m, n = 1, 2, \ldots),$$

and the Fourier series (4), Sec. 27, corresponding to a function f with respect to that orthonormal set, is

$$\frac{1}{\sqrt{2\pi}} \int_{-\pi}^{\pi} f(\xi) \frac{d\xi}{\sqrt{2\pi}} + \sum_{n=1}^{\infty} \left[\frac{\cos nx}{\sqrt{\pi}} \int_{-\pi}^{\pi} f(\xi) \frac{\cos n\xi}{\sqrt{\pi}} d\xi \right.$$
$$\left. + \frac{\sin nx}{\sqrt{\pi}} \int_{-\pi}^{\pi} f(\xi) \frac{\sin n\xi}{\sqrt{\pi}} d\xi \right].$$

This is series (1) with coefficients (2).

The correspondence can be written in the form

$$(5) \quad f(x) \sim \frac{1}{2\pi} \int_{-\pi}^{\pi} f(\xi) \, d\xi + \frac{1}{\pi} \sum_{n=1}^{\infty} \left[\cos nx \int_{-\pi}^{\pi} f(\xi) \cos n\xi \, d\xi \right.$$

$$\left. + \sin nx \int_{-\pi}^{\pi} f(\xi) \sin n\xi \, d\xi \right].$$

A more compact form is

$$(6) \quad f(x) \sim \frac{1}{2\pi} \int_{-\pi}^{\pi} f(\xi) \, d\xi + \frac{1}{\pi} \sum_{n=1}^{\infty} \int_{-\pi}^{\pi} f(\xi) \cos n(x - \xi) \, d\xi.$$

Note that *the constant term* in the series, or the term $\frac{1}{2}a_0$ in series (1), *is the mean value of $f(x)$ over the interval $(-\pi,\pi)$*.

Each term of series (1) is periodic in x with period 2π. Consequently when the series converges to $f(x)$ on the fundamental interval $(-\pi,\pi)$, it converges to a periodic function of period 2π which coincides with f on the fundamental interval; that is, the series represents the *periodic extension* of f for all x. In case f is defined for all x as a periodic function of period 2π, that is, if $f(x + 2\pi) = f(x)$, the series represents f everywhere when the representation is valid on the fundamental interval. Moreover, in that case any interval of length 2π, such as the interval $(0,2\pi)$, serves as a fundamental interval.

Thus *the Fourier series* (1) *serves either of two important purposes:* (*a*) to represent a function defined over the interval $(-\pi,\pi)$, for values of x on that interval, or (*b*) to represent a periodic function, with period 2π, for all values of x. Clearly, it cannot represent a function for all x if the function is not periodic.

In this chapter we shall establish the convergence of the basic Fourier series to $f(x)$ under fairly general conditions on f. Representations by Fourier series on fundamental intervals other than $(-\pi,\pi)$ will follow easily.

36. Example.   Let us write the Fourier series corresponding to the function f defined on the interval $-\pi < x < \pi$ as follows:

$$f(x) = \begin{cases} 0 & \text{when } -\pi < x \leq 0, \\ x & \text{when } 0 \leq x < \pi. \end{cases}$$

The graph of the function is indicated by the heavy lines in

Fig. 9. The Fourier coefficients are

$$a_0 = \frac{1}{\pi} \int_{-\pi}^{\pi} f(x) \, dx = \frac{1}{\pi} \left(0 + \int_0^{\pi} x \, dx \right) = \frac{\pi}{2},$$

$$a_n = \frac{1}{\pi} \int_0^{\pi} x \cos nx \, dx = \left. \frac{\cos nx + nx \sin nx}{\pi n^2} \right]_0^{\pi} = \frac{(-1)^n - 1}{\pi n^2},$$

$$b_n = \frac{1}{\pi} \int_0^{\pi} x \sin nx \, dx = \left. \frac{\sin nx - nx \cos nx}{\pi n^2} \right]_0^{\pi} = -\frac{(-1)^n}{n},$$

where $n = 1, 2, \ldots$.   Therefore, on the interval $(-\pi,\pi)$,

$$f(x) \sim \frac{\pi}{4} + \sum_{n=1}^{\infty} \left[\frac{(-1)^n - 1}{\pi n^2} \cos nx - \frac{(-1)^n}{n} \sin nx \right].$$

We shall soon see why this Fourier series does converge to $f(x)$ when $-\pi < x < \pi$. Then it will follow that the series also

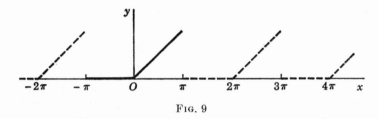

FIG. 9

represents the periodic extension of the function f indicated by the dotted lines in Fig. 9. The periodic function is discontinuous at the points $x = \pm\pi, \pm 3\pi, \ldots$. Our theory will show that the sum of the series at each of those points must be $\pi/2$.

As an indication of the convergence of the series to $f(x)$, a few terms of the series may be summed by composition of ordinates. It will be found, for instance, that the graph of the function

$$y = \frac{\pi}{4} - \frac{2}{\pi} \cos x + \sin x - \frac{1}{2} \sin 2x$$

is a wavy approximation to the graph shown in Fig. 9.

37. Fourier Cosine Series. Sine Series. If $f(-x) = f(x)$ for all values of x for which $f(x)$ is defined, f is an *even* function. Its graph $y = f(x)$ is symmetric to the y axis and, if f is integrable from $x = 0$ to $x = c$, then

$$\int_{-c}^{c} f(x)\ dx = 2 \int_{0}^{c} f(x)\ dx.$$

An *odd* function f is one such that $f(-x) = -f(x)$. Its graph is symmetric to the origin, and its integral from $-c$ to c is zero.

The functions 1, x^2, $\cos kx$, and $x \sin kx$ are even, for example, while x, x^3, and $x^2 \sin kx$ are odd functions. Although most functions are neither even nor odd, each function defined on an interval $(-c,c)$ is expressed as a sum of an even and an odd function by the identity

(1) $$f(x) = \tfrac{1}{2}[f(x) + f(-x)] + \tfrac{1}{2}[f(x) - f(-x)].$$

When f is an even function on the interval $(-\pi,\pi)$, its Fourier coefficients have the values

(2) $$a_n = \frac{2}{\pi} \int_{0}^{\pi} f(x)\ \cos nx\ dx \quad (n = 0, 1, 2, \ldots)$$

and $b_n = 0$ $(n = 1, 2, \ldots)$. Thus its Fourier series reduces to

(3) $$\frac{1}{\pi} \int_{0}^{\pi} f(\xi)\ d\xi + \frac{2}{\pi} \sum_{n=1}^{\infty} \cos nx \int_{0}^{\pi} f(\xi)\ \cos n\xi\ d\xi.$$

This is the *Fourier cosine series* corresponding to f on the interval $(0,\pi)$, with respect to the orthogonal set $\{\cos nx\}$, $n = 0, 1, 2, \ldots$, of eigenfunctions of the Sturm-Liouville problem (compare Problem 2, Sec. 34)

(4) $$X''(x) + \lambda X(x) = 0; \qquad X'(0) = 0, \qquad X'(\pi) = 0.$$

Since the eigenfunctions are even, series (3) represents an even function on the interval $(-\pi,\pi)$ if it converges when $0 < x < \pi$. When f is defined only on the interval $(0,\pi)$, the series is the Fourier series for the even periodic extension of f, with period 2π.

When f is an odd function, $a_n = 0$ and

(5) $$b_n = \frac{2}{\pi} \int_{0}^{\pi} f(x)\ \sin nx\ dx;$$

thus

(6) $$f(x) \sim \frac{2}{\pi} \sum_{n=1}^{\infty} \sin nx \int_{0}^{\pi} f(\xi)\ \sin n\xi\ d\xi.$$

This *Fourier sine series* (compare Sec. 17) will serve to represent odd functions on the interval $(-\pi,\pi)$ or odd periodic functions of period 2π for all x.   But it also serves to represent functions defined only on the interval $(0,\pi)$.   In fact, it is the Fourier series for f with respect to the orthogonal set $\{\sin nx\}$ on the interval

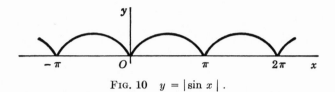

FIG. 10 $y = |\sin x|$.

$(0,\pi)$, the set of eigenfunctions of the Sturm-Liouville problem

(7) $X''(x) + \lambda X(x) = 0;$ $X(0) = 0,$ $X(\pi) = 0.$

Example.   To write the cosine series (3) for the function $f(x) = \sin x$ on the interval $(0,\pi)$, we note that

$$a_0 = \frac{2}{\pi} \int_0^\pi \sin x \, dx = \frac{4}{\pi}, \qquad a_1 = \frac{2}{\pi} \int_0^\pi \sin x \cos x \, dx = 0$$

and, when $n = 2, 3, \ldots$, that

$$a_n = \frac{2}{\pi} \int_0^\pi \sin x \cos nx \, dx$$

$$= \frac{1}{\pi} \int_0^\pi [\sin (1 + n)x + \sin (1 - n)x] \, dx$$

$$= \frac{1}{\pi} \left[\frac{\cos (1 + n)x}{1 + n} + \frac{\cos (1 - n)x}{1 - n} \right]_\pi^0 = \frac{2}{\pi} \frac{1 + (-1)^n}{1 - n^2}.$$

Thus $a_{2n+1} = 0$ and $a_{2n} = 4\pi^{-1}(1 - 4n^2)^{-1}$ so that

(8) $\sin x \sim \dfrac{2}{\pi} - \dfrac{4}{\pi} \sum_{n=1}^{\infty} \dfrac{\cos 2nx}{4n^2 - 1}$ $(0 \leqq x \leqq \pi).$

Let us assume that the correspondence here is an equality for each value of x on the interval indicated as we shall show later. Then for all values of x outside that interval the series converges to the even periodic extension of $\sin x$ of period 2π.   That extension, shown in Fig. 10, is the function $|\sin x|$.

Since $\sin x$ is orthogonal to $\sin nx$, when $n = 2, 3, \ldots$, on the interval $(0,\pi)$, the Fourier sine series (6) for the function $\sin x$ consists of the single term $\sin x$.

38. One-sided Derivatives. Let f be a function whose limit from the right, $f(x_0 + 0)$, exists at a point x_0 (Sec. 26). The *right-hand derivative* of f, or the *derivative from the right*, at that point is defined as follows:

$$(1) \qquad f'_R(x_0) = \lim_{\epsilon \to 0} \frac{f(x_0 + \epsilon) - f(x_0 + 0)}{\epsilon} \qquad (\epsilon > 0),$$

when the limit exists. Note that $f(x_0 + 0)$ must exist if $f'_R(x_0)$ is to exist. In case the ordinary (two-sided) derivative $f'(x_0)$ exists, then $f'_R(x_0) = f'(x_0)$.

Similarly, when $f(x_0 - 0)$ exists, the *left-hand derivative* of f at x_0 is defined by the formula

$$(2) \qquad f'_L(x_0) = \lim_{\epsilon \to 0} \frac{f(x_0 - 0) - f(x_0 - \epsilon)}{\epsilon} \qquad (\epsilon > 0),$$

when the limit exists, and $f'_L(x_0) = f'(x_0)$ if $f'(x_0)$ exists.

To illustrate, consider first the continuous function

$$(3) \qquad f(x) = \begin{cases} x^2 & \text{when } x \leqq 0, \\ \sin x & \text{when } x \geqq 0. \end{cases}$$

We find that $f'_R(0) = 1$ and $f'_L(0) = 0$ as the graph of f indicates; $f'(0)$ does not exist. For the step function

$$h(x) = \begin{cases} 0 & \text{when } x < 0, \\ 1 & \text{when } x > 0, \end{cases}$$

$h'(0)$ does not exist; but both $h'_R(0)$ and $h'_L(0)$ exist and have the common value zero. The function $\sqrt{x}$ $(x \geqq 0)$ is an example of a continuous function that has no right-hand derivative at the point $x = 0$.

If each of two functions f and g has a right-hand derivative at a point x_0, so does their product. A direct proof is left to the problems. But a proof can be based on the corresponding property of ordinary derivatives. We use $f(x_0 + 0)$ and $g(x_0 + 0)$ as the values of f and g at x_0; we also define those functions when $x \leqq x_0$ as the linear functions represented by the tangent lines with slopes $f'_R(x_0)$ and $g'_R(x_0)$, respectively. Those continuations of f and g are differentiable at x_0, with derivatives equal to the right-hand derivatives. Thus the derivative of their product exists there; its value is the right-hand derivative of $f(x)g(x)$ at x_0.

Likewise, if $f_L'(x_0)$ and $g_L'(x_0)$ exist, the left-hand derivative of the product $f(x)g(x)$ exists at x_0.

A further property of one-sided derivatives will be useful in our theory of Fourier series and integrals.

Suppose that both f and f' are sectionally continuous functions on some interval, and let the interval $a \leqq x \leqq b$ denote any one of the subintervals interior to which f and f' are continuous and have one-sided limits from the interior at the end points. If we define $f(a)$ as $f(a + 0)$ and $f(b)$ as $f(b - 0)$, then f is continuous on the closed interval $a \leqq x \leqq b$. Also f' exists on the open interval, so the law of the mean applies; that is, for each number λ $(0 < \lambda < b - a)$ a number θ $(0 < \theta < 1)$ exists such that

$$(4) \qquad \frac{f(a + \lambda) - f(a + 0)}{\lambda} = f'(a + \theta\lambda).$$

Since $f'(a + 0)$ exists, the limit of $f'(a + \theta\lambda)$, as $\lambda \to 0$, exists and has that value. The difference quotient on the left therefore has the same limit; that is, $f_R'(a) = f'(a + 0)$. Similarly, $f_L'(b) = f'(b - 0)$.

Thus *at each point x_0 of a closed interval on which both f and f' are sectionally continuous the one-sided derivatives of f, from the interior of the interval, exist and are the same as the corresponding one-sided limits of f':*

$$(5) \qquad f_R'(x_0) = f'(x_0 + 0), \qquad f_L'(x_0) = f'(x_0 - 0).$$

The continuous function

$$(6) \qquad f(x) = x^2 \sin \frac{1}{x} \qquad \text{when } x \neq 0,$$
$$f(0) = 0,$$

illustrates the distinction between one-sided derivatives and one-sided limits of derivatives. Here $f_R'(0) = f_L'(0) = 0$, while the one-sided limits $f'(+0)$ and $f'(-0)$ do not exist. The proof is left as a problem.

PROBLEMS

1. Verify directly by integration that the set (4), Sec. 35, is orthonormal on the interval $(-\pi, \pi)$.

Write the Fourier series on the interval $(-\pi,\pi)$ for the functions described in Problems 2 to 5 below.

2. $f(x) = x$ when $-\pi < x < \pi$. Also, note the sum of the series when $x = \pm\pi$.

$$Ans.\ 2 \sum_{n=1}^{\infty} \frac{(-1)^{n+1}}{n} \sin nx.$$

3. $f(x) = \begin{cases} 1 \text{ when } -\pi < x < 0, \\ 2 \text{ when } 0 < x < \pi. \end{cases}$ $Ans.\ \dfrac{3}{2} + \dfrac{2}{\pi} \displaystyle\sum_{n=1}^{\infty} \dfrac{\sin (2n-1)x}{2n-1}.$

4. $f(x) = 0$ when $-\pi \leqq x \leqq 0$, $f(x) = \sin x$ when $0 \leqq x \leqq \pi$. Also, given that the series converges to $f(x)$ when $-\pi \leqq x \leqq \pi$, show graphically the function represented by the series for all x.

$$Ans.\ \frac{1}{\pi} + \frac{1}{2} \sin x - \frac{2}{\pi} \sum_{n=1}^{\infty} \frac{\cos 2nx}{4n^2 - 1}.$$

5. $f(x) = e^x$ when $-\pi < x < \pi$.

$$Ans.\ \frac{2 \sinh \pi}{\pi} \left[\frac{1}{2} + \sum_{n=1}^{\infty} \frac{(-1)^n}{1 + n^2} (\cos nx - n \sin nx) \right].$$

6. Write the Fourier cosine series for the function $f(x) = x$ on the interval $0 \leqq x \leqq \pi$. Given that the series represents $f(x)$ on that interval, show graphically the function represented by the series for all x.

$$Ans.\ \frac{\pi}{2} - \frac{4}{\pi} \sum_{n=1}^{\infty} \frac{\cos (2n-1)x}{(2n-1)^2}.$$

7. Prove that the cosine series found in Problem 6 converges uniformly with respect to x for all x and hence that the series does represent some even periodic function which is everywhere continuous (see Sec. 14).

Write (a) the Fourier cosine series, and (b) the Fourier sine series, corresponding to the functions described on the interval $(0,\pi)$ in Problems 8 to 10:

8. $f(x) = 1$ $(0 < x < \pi)$. $Ans.\ (a)\ 1;\ (b)\ \dfrac{4}{\pi} \displaystyle\sum_{n=1}^{\infty} \dfrac{\sin (2n-1)x}{2n-1}.$

9. $f(x) = \pi - x$ $(0 < x < \pi)$.

$$Ans.\ (a)\ \frac{\pi}{2} + \frac{4}{\pi} \sum_{n=1}^{\infty} \frac{\cos (2n-1)x}{(2n-1)^2};\ (b)\ 2 \sum_{n=1}^{\infty} \frac{\sin nx}{n}.$$

10. $f(x) = 1$ when $0 < x < \pi/2$, $f(x) = 0$ when $\pi/2 < x < \pi$.

$$Ans. \ (b) \ \frac{2}{\pi} \sum_{n=1}^{\infty} \left(1 - \cos \frac{n\pi}{2} \right) \frac{\sin nx}{n}.$$

11. For the function f defined by formula (6), Sec. 38, prove that $f'_R(0) = 0$ and that $f'(+0)$ does not exist.

12. Given that the right-hand derivatives of two functions f and g exist at a point x_0, prove that the product of those functions has a right-hand derivative there by inserting the term $f(x_0 + \epsilon)g(x_0 + 0)$ and its negative in the difference

$$\Delta(fg) = f(x_0 + \epsilon)g(x_0 + \epsilon) - f(x_0 + 0)g(x_0 + 0)$$

and taking the limit of $\Delta(fg)/\epsilon$ as $\epsilon \to +0$.

39. An Integration Formula.   We now show that

$$(1) \qquad\qquad \int_0^\infty \frac{\sin x}{x} \, dx = \frac{\pi}{2}.$$

This formula will be used in the following section.

The even function $S(x) = x^{-1} \sin x$ $(x \neq 0)$, $S(0) = 1$ is defined for all x by the alternating series

$$(2) \qquad\qquad S(x) = 1 - \frac{x^2}{3!} + \frac{x^4}{5!} - \frac{x^6}{7!} + \cdots.$$

Therefore $0 < S(x) < 1$ when $0 < x \leq 1$. Also $|S(x)| < |\sin x|$ when $x > 1$. Hence $|S(x)| \leq 1$ when $x \geq 0$.

On the interval $0 < x < \pi$, $\sin x > 0$; when $\pi < x < 2\pi$, $\sin x < 0$, etc. Thus if $x_0 > 0$ and n is the greatest integer such that $n\pi \leq x_0$,

$$(3) \quad \int_0^{x_0} S \, dx = \int_0^\pi S \, dx + \int_\pi^{2\pi} S \, dx + \cdots + \int_{(n-1)\pi}^{n\pi} S \, dx$$
$$+ \int_{n\pi}^{x_0} S \, dx$$
$$= A_0 - A_1 + A_2 - \cdots + (-1)^{n-1} A_{n-1}$$
$$+ (-1)^n \theta_n A_n,$$

where A_k denotes the absolute value of the integral of S from $k\pi$ to $(k + 1)\pi$ and $0 \leq \theta_n < 1$. Note that $A_0 < \pi$. On the kth interval $(k > 0)$, $|S(x)| < 1/(k\pi)$ and hence $A_k < 1/k$. Also, the function S is such that $A_{k+1} < A_k$. The alternating series

$$\lim_{n \to \infty} \sum_{k=0}^{n-1} (-1)^k A_k$$

therefore converges. Since $\theta_n A_n \to 0$ as $n \to \infty$ and $n \to \infty$ when $x_0 \to \infty$, the limit of the integral (3), as $x_0 \to \infty$, exists. That is, the improper integral (1) exists.

That integral has the value $F(0)$, where

$$(4) \qquad\qquad F(t) = \int_0^\infty e^{-tx} \frac{\sin x}{x} \, dx \qquad\qquad (t \geqq 0).$$

The integral (4) converges when $t > 0$ because the absolute value of its integrand does not exceed $\exp(-tx)$; also,

$$(5) \qquad\qquad |F(t)| \leqq \int_0^\infty e^{-tx} \, dx = \frac{1}{t} \qquad\qquad (t > 0).$$

To see that F is continuous when $t \geqq 0$ we first note that, since the integral exists, we can write

$$(6) \quad F(t) = \lim_{n \to \infty} \sum_{k=0}^n (-1)^k \int_{k\pi}^{(k+1)\pi} (-1)^k e^{-tx} \frac{\sin x}{x} \, dx \quad (t \geqq 0).$$

Each integral here has a positive value not greater than A_k in equation (3). Thus the absolute value of the remainder in this alternating series after n terms does not exceed A_n, and $A_n < 1/n$ independent of t. The uniform convergence of the series and the continuity of F follow readily. In particular, $F(+0) = F(0)$.

Now when $t > 0$,

$$(7) \qquad\qquad F'(t) = -\int_0^\infty e^{-tx} \sin x \, dx$$

because this integral is uniformly convergent on the interval $t \geqq t_0$ where t_0 is any positive number. Thus $F'(t) = -(t^2 + 1)^{-1}$ and

$$(8) \qquad\qquad F(t) = -\arctan t + C.$$

The inequality (5) shows that $F(t) \to 0$ when $t \to \infty$. Therefore $C = \pi/2$. But $C = F(+0) = F(0)$, so formula (1) is established.

40. Preliminary Theory. In order to establish the convergence of a Fourier series to its function a few preliminary theorems, or lemmas, on limits of trigonometric integrals are useful. The lemmas will be so formulated that they can be used as well in the theory of Fourier integrals, where it is essential that the parameter k in the lemmas be permitted to vary continuously rather than through the integers only. In the latter case

$(k = n)$ the limits in Lemma 1 follow easily from equation (6), Sec. 28.   Lemma 1 is one form of the *Riemann-Lebesgue theorem.*

Lemma 1. *If a function F is sectionally continuous on an interval $a \leqq x \leqq b$, then*

$$(1) \qquad \lim_{k \to \infty} \int_a^b F(x) \sin kx \, dx = 0,$$

$$(2) \qquad \lim_{k \to \infty} \int_a^b F(x) \cos kx \, dx = 0.$$

The interval (a,b) can be divided into a finite number of intervals on each of which F is continuous even at the end points if we simply use the limits from the interior as the values of F at those points.   Let any one of those intervals be denoted by $p \leqq x \leqq q$.   Then formula (1) will follow if it is shown that

$$(3) \qquad \lim_{k \to \infty} \int_p^q F(x) \sin kx \, dx = 0.$$

Divide the interval (p,q) into n equal parts by the points $x_0 = p, x_1, x_2, \ldots, x_n = q$.   The integral in equation (3) can be written

$$\sum_{i=0}^{n-1} \int_{x_i}^{x_{i+1}} F(x) \sin kx \, dx$$

or $\quad \displaystyle\sum_{i=0}^{n-1} \left\{ F(x_i) \int_{x_i}^{x_{i+1}} \sin kx \, dx + \int_{x_i}^{x_{i+1}} [F(x) - F(x_i)] \sin kx \, dx \right\}.$

We carry out the first integration here.   The absolute value of the second integral does not exceed the integral of $|F(x) - F(x_i)|$. Therefore

$$(4) \quad \left| \int_p^q F(x) \sin kx \, dx \right| \leqq \sum_{i=0}^{n-1} \left| F(x_i) \frac{\cos kx_i - \cos kx_{i+1}}{k} \right|$$

$$+ \sum_{i=0}^{n-1} \int_{x_i}^{x_{i+1}} |F(x) - F(x_i)| \, dx.$$

If M denotes the maximum value of $|F(x)|$ on the interval $p \leqq x \leqq q$, the first sum on the right does not exceed the number

$$(5) \qquad \sum_{i=0}^{n-1} \frac{2M}{k} = \frac{2Mn}{k}.$$

To establish the limit (3) we shall prove that, for each positive number ϵ, a number k_ϵ exists such that

$$
(6) \qquad\qquad \left| \int_p^q F(x) \sin kx \, dx \right| < \epsilon \qquad \text{whenever } k > k_\epsilon.
$$

We have introduced the n equal subdivisions of the interval only as a means of making the proof.

The function F is continuous on the closed interval $p \leqq x \leqq q$. It is therefore uniformly continuous on that interval; that is, to each positive number ϵ' there corresponds a number δ such that

$$
\left| F(x) - F(x_i) \right| < \epsilon' \qquad \text{whenever } |x - x_i| < \delta,
$$

where x and x_i are points on that interval.[1] Let ϵ be the number used in condition (6). We write $\epsilon' = \frac{1}{2}\epsilon/(q - p)$. Then if $n = n_\epsilon$, where the integer n_ϵ is taken large enough that the length of each subinterval (x_i, x_{i+1}) is less than δ,

$$
\sum_{i=0}^{n-1} \int_{x_i}^{x_{i+1}} |F(x) - F(x_i)| \, dx < \frac{\epsilon}{2(q-p)} \sum_{i=0}^{n-1} (x_{i+1} - x_i) = \frac{\epsilon}{2}
$$

$$
(n = n_\epsilon).
$$

It follows from conditions (4) and (5) that

$$
(7) \qquad \left| \int_p^q F(x) \sin kx \, dx \right| < \frac{2M}{k} n_\epsilon + \frac{\epsilon}{2} \qquad \text{when } \frac{q - p}{n_\epsilon} < \delta.
$$

For this fixed value n_ϵ, we now select k large enough that $2M n_\epsilon / k < \epsilon/2$, say $k > k_\epsilon$ where $k_\epsilon = 4M n_\epsilon / \epsilon$. Then condition (6) is satisfied and the limit (1) is established.

The modifications needed when $\sin kx$ is replaced by $\cos kx$, to prove formula (2), are elementary.

Lemma 2. *If F is sectionally continuous on an interval $0 \leqq x \leqq b$ and has a right-hand derivative $F_R'(0)$ then*

$$
(8) \qquad\qquad \lim_{k \to \infty} \int_0^b F(x) \frac{\sin kx}{x} \, dx = \frac{\pi}{2} F(+0).
$$

The integral here can be written as the sum

$$
(9) \qquad F(+0) \int_0^b \frac{\sin kx}{x} \, dx + \int_0^b \frac{F(x) - F(+0)}{x} \sin kx \, dx.
$$

---

[1] See P. Franklin, "Treatise on Advanced Calculus," p. 42, 1940; or D. V. Widder, "Advanced Calculus," p. 172, 1961.

We found in Sec. 39 that $\int_0^\infty t^{-1} \sin t \, dt = \pi/2$. Hence

$$\lim_{k \to \infty} \int_0^b \frac{\sin kx}{x} \, dx = \lim_{k \to \infty} \int_0^{kb} \frac{\sin t}{t} \, dt = \frac{\pi}{2}.$$

The function $[F(x) - F(+0)]/x$ is sectionally continuous on the interval $(0,b)$ since F itself is, and the limit

$$\lim_{x \to 0} \frac{F(x) - F(+0)}{x} \qquad (x > 0),$$

which represents $F'_R(0)$, exists. According to Lemma 1, the limit of the second integral in expression (9), as $k \to \infty$, is zero. Hence equation (8) follows, and the lemma is proved.

Lemma 3. *Let a function F be sectionally continuous on an interval (a,b) and have derivatives from the right and left at a point x_0, where $a < x_0 < b$. Then*

$$(10) \quad \lim_{k \to \infty} \int_a^b F(x) \frac{\sin k(x - x_0)}{x - x_0} \, dx = \pi \frac{F(x_0 + 0) + F(x_0 - 0)}{2}.$$

The integral here can be written as the sum

$$\int_a^{x_0} F(s) \frac{\sin k(x_0 - s)}{x_0 - s} \, ds + \int_{x_0}^b F(t) \frac{\sin k(t - x_0)}{t - x_0} \, dt.$$

Substituting $x = x_0 - s$ in the first of these integrals and $x = t - x_0$ in the second, we can write their sum as

$$(11) \quad \int_0^{x_0 - a} F(x_0 - x) \frac{\sin kx}{x} \, dx + \int_0^{b - x_0} F(x_0 + x) \frac{\sin kx}{x} \, dx.$$

If we write $G(x) = F(x_0 - x)$ and $H(x) = F(x_0 + x)$, we see that $G(+0) = F(x_0 - 0)$ and $H(+0) = F(x_0 + 0)$. Also, from the definitions of one-sided derivatives we find that $G'_R(0)$ and $H'_R(0)$ have the values $-F'_L(x_0)$ and $F'_R(x_0)$, respectively. The limit of the sum (11), as $k \to \infty$, is therefore given by Lemma 2; it is

$$(12) \quad \frac{\pi}{2} [G(+0) + H(+0)] = \frac{\pi}{2} [F(x_0 - 0) + F(x_0 + 0)],$$

so Lemma 3 is established.

41. A Fourier Theorem. A theorem which gives conditions under which a Fourier series converges to its function is called a *Fourier theorem*. One such theorem will now be established.

The conditions are only sufficient for the representation. It will be convenient to consider the periodic extension of the function.

Theorem 1. *Let f denote a function that is sectionally continuous on the interval* $(-\pi,\pi)$*, and periodic with period* 2π*. Then its Fourier series*

$$(1) \qquad \frac{1}{2\pi} \int_{-\pi}^{\pi} f(\xi)\, d\xi + \frac{1}{\pi} \sum_{n=1}^{\infty} \int_{-\pi}^{\pi} f(\xi)\, \cos n(\xi - x)\, d\xi$$

converges to the value

$$(2) \qquad \tfrac{1}{2}[f(x+0) + f(x-0)] \qquad (-\infty < x < \infty)$$

at every point x where f has a right- and left-hand derivative.

Note that series (1) is the compact form of the basic series (1) in Sec. 35 and that the coefficients a_n and b_n exist in view of the sectional continuity of f. The quantity (2) is the mean value of the limits of the function, from the right and left, at point x; it is simply $f(x)$ if f is continuous at the point.

The sum $S_n(x)$ of the first $n+1$ terms of series (1) can be written

$$S_n(x) = \frac{1}{\pi} \int_{-\pi}^{\pi} f(\xi) \left[\frac{1}{2} + \sum_{m=1}^{n} \cos m(\xi - x) \right] d\xi.$$

We apply Lagrange's trigonometric identity (Sec. 30) to the sum in brackets to write

$$(3) \qquad S_n(x) = \frac{1}{\pi} \int_{-\pi}^{\pi} f(\xi)\, \frac{\sin\left[(n + \frac{1}{2})(\xi - x)\right]}{2 \sin\left[\frac{1}{2}(\xi - x)\right]}\, d\xi.$$

It is easy to verify that the integrand here is a periodic function of ξ with period 2π; hence its integral over every interval of length 2π is the same. Let us integrate over the interval $(a, a + 2\pi)$, where the number a is selected so that the point x is interior to that interval; that is, $a < x < a + 2\pi$.

Introducing the factor $\xi - x$ in both numerator and denominator of the integrand, we have

$$(4) \qquad S_n(x) = \frac{1}{\pi} \int_{a}^{a+2\pi} F(\xi)\, \frac{\sin\left[(n + \frac{1}{2})(\xi - x)\right]}{\xi - x}\, d\xi,$$

where, for the fixed value of x $(a < x < a + 2\pi)$,

$$(5) \qquad\qquad F(\xi) = f(\xi)\, \frac{\frac{1}{2}(\xi - x)}{\sin\left[\frac{1}{2}(\xi - x)\right]}.$$

Points x and ξ lie on the interval $a \leq x \leq a + 2\pi$, and x is an interior point. Therefore the distance $|\xi - x|$ between the two points has a maximum value $2D$ that is less than the length 2π of the interval. That is, $\frac{1}{2}|\xi - x| \leq D < \pi$, where D depends on x.

In terms of the function S (Sec. 39), where $S(t) = t^{-1} \sin t$ when $t \neq 0$ and $S(0) = 1$, equation (5) can be written

$$(6) \qquad\qquad F(\xi) = f(\xi)\left[S\left(\frac{\xi - x}{2}\right)\right]^{-1} \qquad (\tfrac{1}{2}|\xi - x| \leq D).$$

Now $S(t) \neq 0$ when $|t| \leq D < \pi$. Since

$$S(t) = 1 - \frac{t^2}{3!} + \frac{t^4}{5!} - \cdots,$$

then $S'(t)$ exists for all t, and it follows that the derivative of $[S(t)]^{-1}$ exists when $|t| < D$. Therefore the function

$$\left[S\left(\frac{\xi - x}{2}\right)\right]^{-1}$$

is continuous and differentiable with respect to ξ when $\frac{1}{2}|\xi - x| < D$. In view of equation (6), $F(\xi)$ must be sectionally continuous on the interval $(a, a + 2\pi)$ and both its one-sided derivatives must exist at the point $\xi = x$ because F is the product of two functions that have those properties. Moreover,

$$F(x + 0) = f(x + 0), \qquad F(x - 0) = f(x - 0).$$

Therefore F satisfies the conditions in Lemma 3 on the interval $a \leq \xi \leq a + 2\pi$. When we apply that lemma to the integral in equation (4), where $n + \frac{1}{2} = k$ and x serves as x_0, we see that

$$\lim_{n \to \infty} S_n(x) = \frac{F(x + 0) + F(x - 0)}{2} = \frac{f(x + 0) + f(x - 0)}{2}.$$

This is the same as the statement in the theorem.

42. Discussion of the Theorem. Suppose the function f is defined only on the interval $(-\pi, \pi)$ and that it is sectionally continuous there. Then Theorem 1 applies to the periodic extension of f.   Thus at each interior point x of the interval where both one-sided derivatives exist, the Fourier series for f converges to the mean value

$$\tfrac{1}{2}[f(x + 0) + f(x - 0)] \qquad (-\pi < x < \pi).$$

But *at both end points* $x = \pm\pi$ *it converges to the value*

(1) $$\tfrac{1}{2}[f(\pi - 0) + f(-\pi + 0)],$$

provided $f'_R(-\pi)$ and $f'_L(\pi)$ exist because this is the mean value of the one-sided limits of the periodic function at each point.   The mean value (1) reduces to $f(\pi - 0)$, or to $f(-\pi + 0)$, if and only if $f(\pi - 0) = f(-\pi + 0)$.

Note that the theorem assumes the existence of both one-sided derivatives of the function only at those points where it ensures convergence of the series to the mean value of the function. The function $g(x) = x^{\frac{2}{3}}$, for instance, is continuous on the interval $-\pi \leq x \leq \pi$, and it has one-sided derivatives there except at the point $x = 0$; also $g(-\pi) = g(\pi)$.   Hence Theorem 1 shows that the Fourier series for g does converge to $g(x)$ when $-\pi \leq x < 0$ and when $0 < x \leq \pi$; it does not ensure convergence when $x = 0$.

In case *both f and f' are sectionally continuous* on the interval $(-\pi,\pi)$, then the one-sided derivatives of the periodic extension of f exist at every point, so *the series converges everywhere* to the mean value of the limits from the right and left for the periodic extension.

In Sec. 36 we wrote the Fourier series for the function

(2) $$f(x) = \begin{cases} 0 & \text{when } -\pi < x \leq 0, \\ x & \text{when } 0 \leq x < \pi. \end{cases}$$

This function is continuous $(-\pi < x < \pi)$.   Both f and f' are sectionally continuous on the interval $(-\pi,\pi)$, so the one-sided derivatives of the periodic extension of f, with period 2π, exist everywhere.   Also, $f(-\pi + 0) = 0$ and $f(\pi - 0) = \pi$.   Theorem 1 therefore shows that the Fourier series found in Sec. 36 converges to $f(x)$ when $-\pi < x < \pi$ and to $\tfrac{1}{2}(0 + \pi)$ or $\pi/2$ when $x = \pm\pi$.   The periodic graph in Fig. 9, with the addition of the points $(\pm\pi,\pi/2)$, $(\pm 3\pi,\pi/2)$, . . . , represents the sum of the series for all x.

Representations of sectionally continuous functions on the interval $(0,\pi)$ by their Fourier cosine series and their Fourier sine series are special cases of Theorem 1 because the theorem applies to the even and odd extensions of the functions onto the interval $(-\pi,\pi)$. The series for those extensions reduce to the cosine series and the sine series, respectively (Sec. 37). We state the special Fourier theorems as a corollary.

Corollary 1. *Let f be a sectionally continuous function on the interval $(0,\pi)$, defined, for convenience, at each interior point x_0 where f is discontinuous, as the mean value $\frac{1}{2}[f(x_0 + 0) + f(x_0 - 0)]$. Then at each point x $(0 < x < \pi)$ where $f'_R(x)$ and $f'_L(x)$ exist, f is represented by its Fourier cosine series*

$$(3) \qquad f(x) = \frac{1}{2} a_0 + \sum_{n=1}^{\infty} a_n \cos nx \qquad (0 < x < \pi),$$

where

$$(4) \qquad a_n = \frac{2}{\pi} \int_0^\pi f(x) \cos nx \, dx \quad (n = 0, 1, 2, \ldots);$$

also by its Fourier sine series

$$(5) \qquad f(x) = \sum_{n=1}^{\infty} b_n \sin nx \qquad (0 < x < \pi),$$

where

$$(6) \qquad b_n = \frac{2}{\pi} \int_0^\pi f(x) \sin nx \, dx \qquad (n = 1, 2, \ldots).$$

In view of the even periodic function represented by the cosine series (3), that series converges at the point $x = 0$ to $f(+0)$ if $f'_R(0)$ exists; at $x = \pi$ it converges to $f(\pi - 0)$ if $f'_L(\pi)$ exists. The sum of the sine series (5) is clearly zero when $x = 0$ and when $x = \pi$.

Broader conditions than those given in Theorem 1, under which a Fourier series converges to its function, are stated in Chap. 5.[1]

PROBLEMS

1. Use Theorem 1 and series (8), Sec. 37, to point out why it is true that, for every value of x,

$$|\sin x| = \frac{2}{\pi} - \frac{4}{\pi} \sum_{n=1}^{\infty} \frac{\cos 2nx}{4n^2 - 1} \qquad (-\infty < x < \infty).$$

[1] For other introductions to the theory of Fourier series see the references to Bôcher and Jackson in the Bibliography. Our proof of Theorem 1 originated from Bôcher's treatment.

Note that these two summations follow as a consequence, by writing $x = 0$ and $x = \pi/2$:

$$\sum_{n=1}^{\infty} \frac{1}{4n^2 - 1} = \frac{1}{2}; \qquad \sum_{n=1}^{\infty} \frac{(-1)^n}{4n^2 - 1} = \frac{1}{2} - \frac{\pi}{4}.$$

2. With the aid of Theorem 1, state why the series found in Problem 4, Sec. 38, for the function $f(x) = 0$ $(-\pi \leqq x \leqq 0)$, $f(x) = \sin x$ $(0 \leqq x \leqq \pi)$ must converge to $f(x)$ everywhere on the interval

$$-\pi \leqq x \leqq \pi.$$

3. Show that each of the functions described in Problems 3 and 5, Sec. 38, satisfies conditions under which the series found there must converge to the value of the function except at certain points on the interval $-\pi \leqq x \leqq \pi$, and give the sum of the series at those points.

Ans. Prob. 3: $x = 0$, $\pm\pi$, sum $= \frac{3}{2}$.   Prob. 5: $x = \pm\pi$, sum $= \cosh \pi$.

4. State why the function $|x|$ is represented by its Fourier series everywhere on the interval $-\pi \leqq x \leqq \pi$. From that series, found in Problem 6, Sec. 38, show that

$$\sum_{n=1}^{\infty} \frac{1}{(2n - 1)^2} = \frac{\pi^2}{8}.$$

5. Describe graphically the function to which the sine series found in Problem 10, Sec. 38, must converge for all values of x $(-\infty < x < \infty)$.

6. Show that the odd function $f(x) = x^{\frac{1}{3}}$ $(-\pi < x < \pi)$ does not have one-sided derivatives at the point $x = 0$. Without finding its Fourier series, state why that series does converge to $f(x)$ on the interval $-\pi < x < \pi$, including the point $x = 0$, thus illustrating the fact that the existence of one-sided derivatives is not a necessary condition for convergence.

43. Other Forms of Fourier Series.   Let c denote any positive number and f a periodic function of period $2c$ that is sectionally continuous over the interval $(-c,c)$. We define $f(x)$ at each point of discontinuity as the mean value of $f(x + 0)$ and $f(x - 0)$. A Fourier theorem for f follows from Theorem 1 by changing the unit of length on the x axis.

We introduce a new independent variable t,

(1) $$t = \frac{\pi}{c} x;$$

then $-\pi < t < \pi$ when $-c < x < c$, and $f(x) = f(ct/\pi)$. We write $F(t) = f(ct/\pi)$. The function F is periodic with period 2π; and continuous at a point t if f is continuous at the corresponding point x. From the sectional continuity of f on the interval $(-c,c)$ we can conclude that F is sectionally continuous on $(-\pi,\pi)$, and from the definition of $f(x)$ at points of discontinuity we find that $F(t)$ has the mean of the values $F(t + 0)$ and $F(t - 0)$ at its points of discontinuity. Moreover, the existence of one-sided derivatives of f at a point x ensures the existence of those derivatives of F at the corresponding point t.

According to Theorem 1, at each point t where $F'_R(t)$ and $F'_L(t)$ exist, $F(t)$ is represented by its Fourier series; that is,

$$(2) \qquad f\left(\frac{ct}{\pi}\right) = \frac{1}{2} a_0 + \sum_{n=1}^{\infty} (a_n \cos nt + b_n \sin nt),$$

where

$$(3) \qquad \begin{aligned} a_n &= \frac{1}{\pi} \int_{-\pi}^{\pi} f\left(\frac{ct}{\pi}\right) \cos nt \, dt, \\ b_n &= \frac{1}{\pi} \int_{-\pi}^{\pi} f\left(\frac{ct}{\pi}\right) \sin nt \, dt. \end{aligned}$$

Under the substitution (1), equation (2) becomes

$$(4) \qquad f(x) = \frac{1}{2} a_0 + \sum_{n=1}^{\infty} \left(a_n \cos \frac{n\pi x}{c} + b_n \sin \frac{n\pi x}{c} \right).$$

Formulas (3) for the coefficients can be written, by making the substitution (1) for the variable of integration,

$$(5) \qquad \begin{aligned} a_n &= \frac{1}{c} \int_{-c}^{c} f(x) \cos \frac{n\pi x}{c} \, dx \quad (n = 0, 1, 2, \ldots), \\ b_n &= \frac{1}{c} \int_{-c}^{c} f(x) \sin \frac{n\pi x}{c} \, dx \qquad (n = 1, 2, \ldots). \end{aligned}$$

The series in equation (4) with coefficients (5) is *the Fourier series for periodic functions of period 2c.* The set of functions

$$(6) \qquad \left\{ 1, \cos \frac{m\pi x}{c}, \sin \frac{n\pi x}{c} \right\} \quad (m, n = 1, 2, \ldots)$$

used in expansion (4) is the orthogonal set of eigenfunctions of the eigenvalue problem

$$(7) \quad X'' + \lambda X = 0; \qquad X(-c) = X(c), \qquad X'(-c) = X'(c).$$

The representation (4) is used for functions defined only on the interval $(-c,c)$, as well as for periodic functions. But as in the special case $c = \pi$, it is simpler to state the Fourier theorem as follows in terms of periodic extensions of the functions.

Corollary 2. *If a function f is periodic with period 2c, sectionally continuous on the interval $(-c,c)$, and defined as its mean value from the right and left at its points of discontinuity, then at each point x where $f'_R(x)$ and $f'_L(x)$ exist, $f(x)$ has the Fourier series representation (4) with the coefficients (5).*

In case f is an even function, the above expansion reduces to the *Fourier cosine series* representation

$$(8) \qquad f(x) = \frac{1}{2} a_0 + \sum_{n=1}^{\infty} a_n \cos \frac{n\pi x}{c},$$

where

$$(9) \qquad a_n = \frac{2}{c} \int_0^c f(x) \cos \frac{n\pi x}{c}\, dx \quad (n = 0, 1, 2, \ldots).$$

This is the representation of a function f on the interval $(0,c)$ in terms of the eigenfunctions of the Sturm-Liouville problem

$$(10) \quad X''(x) + \lambda X(x) = 0; \qquad X'(0) = 0, \qquad X'(c) = 0.$$

When f is odd, the series in the representation (4) becomes the *Fourier sine series* for the interval $(0,c)$, introduced in Sec. 17.

If we express $\cos (n\pi x/c)$ and $\sin (n\pi x/c)$ in terms of imaginary exponential functions and write

$$(11) \quad \gamma_0 = \tfrac{1}{2} a_0, \qquad \gamma_n = \tfrac{1}{2}(a_n - ib_n), \qquad \gamma_{-n} = \tfrac{1}{2}(a_n + ib_n)$$
$$(n = 1, 2, \ldots),$$

the representation (4) takes the form

$$(12) \qquad f(x) = \lim_{m \to \infty} \sum_{n=-m}^{m} \gamma_n \exp \frac{in\pi x}{c}.$$

This is the *exponential form of the Fourier series* expansion of a periodic function with period $2c$. Formula (11) can be written

$$(13) \qquad \gamma_n = \frac{1}{2c} \int_{-c}^{c} f(x) \exp\left(-\frac{in\pi x}{c}\right) dx$$

$$(n = 0, \pm 1, \pm 2, \ldots).$$

Note that the sum in equation (12) includes the term γ_0 corresponding to $n = 0$. The limit there is the *principal value* of the sum

$$(14) \qquad \sum_{n=-\infty}^{\infty} \gamma_n \exp\frac{in\pi x}{c} = \gamma_0 + \sum_{n=1}^{\infty} \gamma_n \exp\frac{in\pi x}{c}$$

$$+ \sum_{n=-1}^{-\infty} \gamma_n \exp\frac{in\pi x}{c},$$

obtained by grouping γ_0 with the first m terms of each of the last two series, then taking the limit, as $m \to \infty$, of the sum of terms in that group. The principal value of a series summed from $n = -\infty$ to $n = \infty$ sometimes exists when the series diverges; but if the series converges, its sum is the same as the principal value.

Details in obtaining form (12) are left to the problems.

44. The Orthonormal Trigonometric Functions. We use the symbol $C'_s(a,b)$ to denote the set, or function space, of all functions f such that f and f' are both sectionally continuous on the interval (a,b). The one-sided derivatives of such functions, from the interior of the interval, therefore exist at every point x $(a \leqq x \leqq b)$; also, the number of discontinuities on the interval is finite.

Corollary 2 applies to the periodic extension, of period $2c$, of each function f of the space $C'_s(-c,c)$, to show that $f(x)$ has the Fourier series representation (4), Sec. 43, at all points x $(-c < x < c)$ where f is continuous. That series is the special case of the generalized Fourier series on the interval $(-c,c)$ with respect to the orthonormal set

$$(1) \qquad \left\{ \frac{1}{\sqrt{2c}}, \frac{1}{\sqrt{c}} \cos\frac{m\pi x}{c}, \frac{1}{\sqrt{c}} \sin\frac{n\pi x}{c} \right\} \quad (m, n = 1, 2, \ldots)$$

on that interval. The functions of this set belong to the space

$C'_s(-c,c)$.   Since the representation is ensured except for at most a finite number of points on the interval, we can state the following result in the terminology introduced in Chap. 3.

Corollary 3. *In the function space* $C'_s(-c,c)$, *the orthonormal set of functions* (1) *is closed in the sense of pointwise convergence. It is also complete.*

It was pointed out in Sec. 29 that the set must be complete if it is closed.   Note that the corollary is stated for functions whose one-sided derivatives exist everywhere on the interval.

From the representations of functions on the interval $(0,c)$ by Fourier cosine series, or by the sine series, corresponding statements follow for the orthonormal set

(2)
$$\left\{ \frac{1}{\sqrt{c}}, \sqrt{\frac{2}{c}} \cos \frac{n\pi x}{c} \right\} \qquad (n = 1, 2, \ldots),$$

and for the orthonormal set

(3)
$$\left\{ \sqrt{\frac{2}{c}} \sin \frac{n\pi x}{c} \right\} \qquad (n = 1, 2, \ldots),$$

on the interval $(0,c)$.

Corollary 4. *In the function space* $C'_s(0,c)$ *each of the orthonormal sets* (2) *and* (3) *is closed in the sense of pointwise convergence, and each is complete.*

PROBLEMS

1. If $f(x + 2c) = f(x)$ for all x and $f(x) = -1$ when $-c < x < 0$, $f(x) = 1$ when $0 < x < c$, and $f(0) = f(c) = 0$, show that for all x $(-\infty < x < \infty)$ it is true that

$$f(x) = \frac{4}{\pi} \sum_{n=1}^{\infty} \frac{1}{2n-1} \sin \frac{(2n-1)\pi x}{c}.$$

2. If $f(x) = 0$ when $-2 < x < 1$ and $f(x) = 1$ when $1 < x < 2$, and $f(1) = f(2) = f(-2) = \frac{1}{2}$, show that, when $-2 \leqq x \leqq 2$,

$$f(x) = \frac{1}{4} - \frac{1}{\pi} \sum_{n=1}^{\infty} \frac{1}{n} \left[\sin \frac{n\pi}{2} \cos \frac{n\pi x}{2} + \left(\cos n\pi - \cos \frac{n\pi}{2} \right) \sin \frac{n\pi x}{2} \right].$$

3. Prove that, when $0 \leqq x \leqq c$,

$$x^2 = \frac{c^2}{3} + \frac{4c^2}{\pi^2} \sum_{n=1}^{\infty} \frac{(-1)^n}{n^2} \cos \frac{n\pi x}{c},$$

hence that $\displaystyle\sum_{n=1}^{\infty} \frac{(-1)^{n+1}}{n^2} = \frac{\pi^2}{12},$ $\displaystyle\sum_{n=1}^{\infty} \frac{1}{n^2} = \frac{\pi^2}{6}.$

4. Show that, when $-1 < x < 1$,

$$x + x^2 = \frac{1}{3} + \frac{2}{\pi} \sum_{n=1}^{\infty} (-1)^n \left(\frac{2}{\pi n^2} \cos n\pi x - \frac{1}{n} \sin n\pi x \right).$$

5. If $f(x) = \cos \pi x$ when $0 < x < 1$ and $f(x) = 0$ when $1 < x < 2$, and $f(0) = \frac{1}{2}$, $f(1) = -\frac{1}{2}$, and $f(x + 2) = f(x)$ for all x, write the Fourier series for f and state why the series must converge everywhere to $f(x)$.

$$Ans. \frac{1}{2} \cos \pi x + \frac{4}{\pi} \sum_{n=1}^{\infty} \frac{n}{4n^2 - 1} \sin 2n\pi x.$$

6. Use the Fourier sine series on the interval $(0,1)$ to show that

$$\cos \pi x = \frac{8}{\pi} \sum_{n=1}^{\infty} \frac{n}{4n^2 - 1} \sin 2n\pi x \qquad (0 < x < 1).$$

7. If $f(x) = \frac{1}{4}c - x$ when $0 \leqq x \leqq \frac{1}{2}c$ and $f(x) = x - \frac{3}{4}c$ when $\frac{1}{2}c \leqq x \leqq c$, use the cosine series on the interval $(0,c)$ to establish the representation

$$f(x) = \frac{2c}{\pi^2} \sum_{n=1}^{\infty} \frac{1}{(2n-1)^2} \cos \frac{(4n-2)\pi x}{c} \qquad (0 \leqq x \leqq c).$$

8. In Sec. 43, write the Fourier representation (4) in the form

$$f(x) = \frac{1}{2} a_0 + \lim_{m \to \infty} \sum_{n=1}^{m} \left(a_n \cos \frac{n\pi x}{c} + b_n \sin \frac{n\pi x}{c} \right);$$

then obtain the exponential form (12) of that representation and formula (13) for the coefficients γ_n. Also, note how those same values of γ_n can be found formally from equation (12) by using the hermitian orthogonality (Sec. 31) of the functions $\exp(in\pi x/c)$ on the interval $(-c,c)$.

9. Use the exponential form (12), Sec. 43, of the Fourier representation, when $c = \pi$, to show that, if $-\pi < x < \pi$,

$$e^x = \frac{\sinh \pi}{\pi} \lim_{m \to \infty} \sum_{n=-m}^{m} \frac{(-1)^n}{1 - in} \exp(inx).$$

(Compare Problem 5, Sec. 38.)

10. State why the set of cosine functions $\{\cos (n\pi x/c)\}$, where $n = 1, 2, \ldots$, excluding a constant function, is closed in the sense of pointwise convergence, in the subspace of the function space $C_s'(0,c)$ consisting of all functions f of that space for which

$$\int_0^c f(x)\ dx = 0.$$

11. If f is sectionally continuous on an interval $(0,c)$ and b_n denotes the coefficients in the Fourier sine series for f on that interval (Sec. 17), use Bessel's inequality (Sec. 28) to prove that the series of terms b_n^2 converges and that

$$\sum_{n=1}^{\infty} b_n^2 \le \frac{2}{c} \int_0^c [f(x)]^2\ dx.$$

Hence, deduce that $b_n \to 0$ and $n \to \infty$, and note that this conclusion also follows from Lemma 1.

12. If f is sectionally continuous on the interval $(0,c)$, prove that the coefficients (9), Sec. 43, satisfy the condition

$$\frac{1}{2} a_0^2 + \sum_{n=1}^{\infty} a_n^2 \le \frac{2}{c} \int_0^c [f(x)]^2\ dx$$

(compare Problem 11), and deduce that $a_n \to 0$ as $n \to \infty$.

13. If f is sectionally continuous on the interval $(-c,c)$, show that the coefficients (5), Sec. 43, satisfy the condition

$$\frac{1}{2} a_0^2 + \sum_{n=1}^{\infty} a_n^2 + \sum_{n=1}^{\infty} b_n^2 \le \frac{1}{c} \int_{-c}^{c} [f(x)]^2\ dx.$$

(Compare Problem 11).

14. (a) The coefficients b_n in the Fourier sine series for a sectionally continuous function f on an interval $(0,c)$ are those for which a finite linear combination of the sine functions becomes the best approximation in the mean to $f(x)$ over the interval $(0,c)$. Show how this follows from Theorem 1, Sec. 28.

(b) Find the values of A_1, A_2, and A_3 such that the function

$$y = A_1 \sin \frac{\pi x}{2} + A_2 \sin \frac{2\pi x}{2} + A_3 \sin \frac{3\pi x}{2}$$

is the best approximation in the mean to the function $f(x) = 1$, over the interval $(0,2)$. Also, draw the graph of y using the coefficients found, and compare it to the graph of $f(x)$.

$Ans.$ $A_1 = 4/\pi$, $A_2 = 0$, $A_3 = 4/(3\pi)$.

15. Theorem 1 ensures the convergence of the Fourier series with corresponding cosine and sine terms grouped as a single term ($a_n \cos nx$ + $b_n \sin nx$). When f satisfies the conditions stated in the theorem and when x is a point such that $f'(x)$ and $f'(-x)$ exist, show that

$$\tfrac{1}{2}[f(x) + f(-x)] = \tfrac{1}{2}a_0 + \sum_{n=1}^{\infty} a_n \cos nx.$$

Thus the cosine terms themselves form a convergent series at the point. Why must the series of sine terms converge there also?

CHAPTER 5

FURTHER PROPERTIES OF FOURIER SERIES

In establishing the uniqueness of solutions of boundary value problems in partial differential equations (Chap. 10) it is sometimes helpful to know conditions on a function under which its Fourier series will converge uniformly over the fundamental interval. We present such conditions in the following section. Differentiation and integration of Fourier series are not involved in our later applications; nevertheless we shall extend our theory here to cover those basic operations on the series. Sections 46 to 49 can be skipped without disrupting the continuity of presentation in the book.

45. Uniform Convergence. If A_n and B_n $(n = 1, 2, \ldots, m)$ represent real numbers, the quadratic equation in x,

$$\sum_{n=1}^{m} (A_n x + B_n)^2 = x^2 \sum_{n=1}^{m} A_n{}^2 + 2x \sum_{n=1}^{m} A_n B_n + \sum_{n=1}^{m} B_n{}^2 = 0,$$

cannot have distinct real roots. For if it has a real root $x = x_0$, then $A_n x_0 + B_n = 0$ for each n. Thus the ratio $-B_n/A_n$ must be independent of n and equal to that number x_0, for each nonzero A_n. In case $A_n = 0$ for some n, then $B_n = 0$. The discriminant of the quadratic equation is therefore negative or zero; that is,

$$(1) \qquad \left(\sum_{n=1}^{m} A_n B_n \right)^2 \leqq \left(\sum_{n=1}^{m} A_n{}^2 \right) \left(\sum_{n=1}^{m} B_n{}^2 \right).$$

Condition (1) is known as *Cauchy's inequality*. When $m = 3$, it simply states that the square of the inner product of two vectors does not exceed the product of the squares of their lengths. The corresponding property for inner products of functions is the

102

Schwarz inequality (Problem 4, Sec. 29).   We shall use Cauchy's inequality in proving the following theorem.

Theorem 1.   *Let* f *be a continuous function on the interval* $-\pi \leq x \leq \pi$ *such that* $f(-\pi) = f(\pi)$, *and let its derivative* f' *be sectionally continuous on that interval.   Then the series*

$$(2) \qquad \sum_{n=1}^{\infty} \sqrt{a_n{}^2 + b_n{}^2}$$

converges, where a_n *and* b_n *are the Fourier coefficients*

$$a_n = \frac{1}{\pi} \int_{-\pi}^{\pi} f(x) \cos nx \, dx, \qquad b_n = \frac{1}{\pi} \int_{-\pi}^{\pi} f(x) \sin nx \, dx.$$

From the comparison test we note that each of the series

$$(3) \qquad \sum_{n=1}^{\infty} |a_n|, \qquad \sum_{n=1}^{\infty} |b_n|$$

converges as a consequence of the convergence of series (2).

The Fourier coefficients of f',

$$(4) \quad \alpha_n = \frac{1}{\pi} \int_{-\pi}^{\pi} f'(x) \cos nx \, dx, \qquad \beta_n = \frac{1}{\pi} \int_{-\pi}^{\pi} f'(x) \sin nx \, dx,$$

exist because of the sectional continuity of f'.   Since f is continuous and $f(-\pi) = f(\pi)$,

$$(5) \qquad \alpha_0 = \frac{1}{\pi} \int_{-\pi}^{\pi} f'(x) \, dx = \frac{1}{\pi} [f(\pi) - f(-\pi)] = 0.$$

Also, when $n = 1, 2, \ldots$, we see by integration by parts that

$$(6) \quad \alpha_n = \frac{n}{\pi} \int_{-\pi}^{\pi} f(x) \sin nx \, dx + \frac{1}{\pi} [f(x) \cos nx]_{-\pi}^{\pi}$$

$$= nb_n + \frac{\cos n\pi}{\pi} [f(\pi) - f(-\pi)] = nb_n,$$

$$(7) \quad \beta_n = -\frac{n}{\pi} \int_{-\pi}^{\pi} f(x) \cos nx \, dx + \frac{1}{\pi} [f(x) \sin nx]_{-\pi}^{\pi} = -na_n.$$

Note that the condition $f(-\pi) = f(\pi)$, under which the periodic extension of f is also continuous, is necessary if equation (6) is to reduce to the form $\alpha_n = nb_n$.

Now let S_m denote the partial sum of the infinite series (2).

In view of relations (6) and (7),

$$S_m = \sum_{n=1}^{m} \sqrt{a_n{}^2 + b_n{}^2} = \sum_{n=1}^{m} \frac{1}{n} \sqrt{\alpha_n{}^2 + \beta_n{}^2}.$$

Since $S_m \geqq 0$, it follows from Cauchy's inequality (1) that

(8)
$$S_m \leqq \left[\sum_{n=1}^{m} \frac{1}{n^2} \sum_{n=1}^{m} (\alpha_n{}^2 + \beta_n{}^2) \right]^{\frac{1}{2}}.$$

The first sum on the right is bounded for all m because the infinite series of positive terms $1/n^2$ converges.

From Bessel's inequality (Sec. 28 or Problem 13, Sec. 44) for the sectionally continuous function f', with respect to the orthonormal set

$$\left\{ \frac{1}{\sqrt{2\pi}}, \frac{1}{\sqrt{\pi}} \cos kx, \frac{1}{\sqrt{\pi}} \sin nx \right\} \quad (k, n = 1, 2, \ldots)$$

on the interval $(-\pi,\pi)$, we find that for every m,

$$\sum_{n=1}^{m} (\alpha_n{}^2 + \beta_n{}^2) \leqq \frac{1}{\pi} \int_{-\pi}^{\pi} [f'(x)]^2 \, dx,$$

since $\alpha_0 = 0$. Therefore the right-hand member of condition (8) is bounded for all m, and so is S_m.

Since S_m is a bounded nondecreasing sequence, its limit as m tends to infinity exists; that is, series (2) converges, as stated in the theorem.

Theorem 2. *Under the conditions stated in Theorem 1 the convergence of the Fourier series*

(9)
$$\tfrac{1}{2}a_0 + \sum_{n=1}^{\infty} (a_n \cos nx + b_n \sin nx)$$

to $f(x)$ on the interval $-\pi \leqq x \leqq \pi$ is absolute and uniform with respect to x on that interval.

The conditions on f and f' ensure the continuity, and existence of one-sided derivatives, of the periodic extension of f for all x. It follows from our Fourier theorem (Sec. 41) that series (9) converges to $f(x)$ everywhere on the interval $-\pi \leqq x \leqq \pi$.   Now

$$|a_n \cos nx + b_n \sin nx| \leqq |a_n| + |b_n|$$

and the series of the constants $|a_n| + |b_n|$ converges because each of series (3) converges.   The comparison test and the Weierstrass M-test (Sec. 14) therefore apply to show that the convergence of series (9) is absolute and uniform as stated in Theorem 2.

The tests apply as well to show the convergence of the series of cosine terms only, or the series of sine terms; in fact, the convergence is absolute and uniform.   Therefore series (9) is the sum of those series; that is,

$$(10) \quad f(x) = \tfrac{1}{2}a_0 + \sum_{n=1}^{\infty} a_n \cos nx + \sum_{n=1}^{\infty} b_n \sin nx$$

$$(-\pi \leqq x \leqq \pi),$$

and both series here converge absolutely and uniformly.

46. Observations.   A Fourier series cannot converge uniformly over an interval that contains a discontinuity of its sum because a uniformly convergent series of continuous functions always converges to a continuous function.   Hence some continuity requirement on f, such as the continuity assumed in Theorem 2, is necessary in order to ensure uniform convergence of the Fourier series to $f(x)$.

Modifications of Theorems 1 and 2 for cosine series or sine series, or for Fourier series on an interval $(-c,c)$, are apparent. For instance, it follows from Theorem 2 that the Fourier cosine series for a continuous function f, on the interval $0 \leqq x \leqq \pi$, converges uniformly to $f(x)$ over that interval if f' is sectionally continuous on the interval.   For the sine series, however, the further conditions $f(0) = f(\pi) = 0$ are needed.

Consider the space of functions satisfying the conditions stated in Theorem 1.   Parseval's equation (Sec. 29) with respect to the orthonormal trigonometric functions on the interval $(-\pi,\pi)$ is satisfied by each function f of that space.   This is seen by multiplying the Fourier series expansion of f by $f(x)$, thus leaving the series still uniformly convergent and integrating, to write

$$\int_{-\pi}^{\pi} [f(x)]^2 \, dx = \tfrac{1}{2}a_0 \int_{-\pi}^{\pi} f(x) \, dx$$

$$+ \sum_{n=1}^{\infty} \left[a_n \int_{-\pi}^{\pi} f(x) \cos nx \, dx + b_n \int_{-\pi}^{\pi} f(x) \sin nx \, dx \right].$$

This is Parseval's equation

$$(1) \qquad \int_{-\pi}^{\pi} [f(x)]^2 \, dx = \pi \left[\tfrac{1}{2}a_0{}^2 + \sum_{n=1}^{\infty} (a_n{}^2 + b_n{}^2) \right].$$

From it we conclude (Sec. 29) that the orthonormal set of trigonometric functions is closed in the sense of mean convergence, in the function space.

47. Differentiation of Fourier Series.   We have seen that the Fourier series for the function $f(x) = x$ on the interval $(-\pi,\pi)$ converges to $f(x)$ at each interior point of the interval; that is,

$$x = 2 \sum_{n=1}^{\infty} (-1)^{n+1} \frac{\sin nx}{n} \qquad (-\pi < x < \pi).$$

But the series here is not differentiable.   The derived series

$$2 \sum_{n=1}^{\infty} (-1)^{n+1} \cos nx$$

does not converge since its nth term fails to approach zero as n tends to infinity.   The periodic extension of f with period 2π, represented by the first series for all x, has discontinuities at the points $x = \pm\pi$.

Continuity of the periodic functions is an important condition for differentiability of Fourier series.   Sufficient conditions can be stated as follows.

Theorem 3.   *Let f be a continuous function on the interval $-\pi \leqq x \leqq \pi$ such that $f(-\pi) = f(\pi)$, and let f' be sectionally continuous on that interval.   Then the Fourier series in the representation*

$$(1) \quad f(x) = \tfrac{1}{2}a_0 + \sum_{n=1}^{\infty} (a_n \cos nx + b_n \sin nx) \quad (-\pi \leqq x \leqq \pi),$$

where $a_n = \dfrac{1}{\pi} \displaystyle\int_{-\pi}^{\pi} f(x) \cos nx \, dx, \quad b_n = \dfrac{1}{\pi} \displaystyle\int_{-\pi}^{\pi} f(x) \sin nx \, dx, \quad$ *is*

differentiable at each point where $f''(x)$ exists:

$$(2) \qquad f'(x) = \sum_{n=1}^{\infty} n(-a_n \sin nx + b_n \cos nx) \quad (-\pi < x < \pi).$$

Since the function f' satisfies the conditions of our Fourier theorem, it is represented by its Fourier series at each point x where its derivative $f''(x)$ exists. At such a point f' is continuous so that

$$(3) \quad f'(x) = \tfrac{1}{2}\alpha_0 + \sum_{n=1}^{\infty} (\alpha_n \cos nx + \beta_n \sin nx) \quad (-\pi < x < \pi),$$

where α_n and β_n are the coefficients (4), Sec. 45. But when f satisfies the conditions stated in the first sentence of Theorem 3, we found in Sec. 45 that

$$(4) \quad \alpha_0 = 0, \quad \alpha_n = nb_n, \quad \beta_n = -na_n \quad (n = 1, 2, \ldots).$$

When these substitutions are made, equation (3) takes the form (2). This completes the proof of the theorem.

At a point x where $f''(x)$ does not exist, but where f' has derivatives from the right and left, the differentiation is still valid in the sense that the series in equation (2) converges to the mean of the values $f'(x + 0)$ and $f'(x - 0)$. That is also true for the periodic extension of f.

Theorem 3 applies with natural changes to other forms of Fourier series. For instance, if f is continuous and f' is sectionally continuous on an interval $0 \leq x \leq c$, then the Fourier cosine series for f on that interval is differentiable at each point where $f''(x)$ exists.

48. Integration of Fourier Series. Integration of a Fourier series is possible under much more general conditions than those for differentiation. This is to be expected because an integration introduces a factor n in the denominator of the general term. It will be shown in the following theorem that it is not even essential that the original series converge to its function in order that the integrated series converge to the integral of the function. Of course, the integrated series is not a Fourier series if $a_0 \neq 0$, for it contains a term $a_0x/2$.

Theorem 4. *Let f be sectionally continuous on the interval $(-\pi,\pi)$. Then whether the Fourier series corresponding to f,*

$$(1) \quad f(x) \sim \tfrac{1}{2}a_0 + \sum_{n=1}^{\infty} (a_n \cos nx + b_n \sin nx),$$

converges or not, the following equality is true:

$$(2) \quad \int_{-\pi}^{x} f(\xi) \, d\xi = \frac{1}{2} a_0(x + \pi)$$

$$+ \sum_{n=1}^{\infty} \frac{1}{n} [a_n \sin nx - b_n (\cos nx - \cos n\pi)],$$

when $-\pi \leqq x \leqq \pi$. *The latter series is obtained by integrating the former one term by term.*

Since f is sectionally continuous, the function F, where

$$(3) \qquad\qquad F(x) = \int_{-\pi}^{x} f(\xi) \, d\xi - \frac{1}{2} a_0 x,$$

is continuous; moreover

$$F'(x) = f(x) - \frac{1}{2} a_0,$$

except at points where f is discontinuous. Therefore F' is sectionally continuous on the interval $(-\pi,\pi)$. Also,

$$F(\pi) = \int_{-\pi}^{\pi} f(\xi) \, d\xi - \frac{1}{2} a_0 \pi = a_0 \pi - \frac{1}{2} a_0 \pi = \frac{1}{2} a_0 \pi,$$

and $F(-\pi) = \frac{1}{2} a_0 \pi$; hence $F(\pi) = F(-\pi)$. According to our Fourier theorem then, for all x on the interval $-\pi \leqq x \leqq \pi$, it is true that

$$F(x) = \frac{1}{2} A_0 + \sum_{n=1}^{\infty} (A_n \cos nx + B_n \sin nx),$$

where

$$A_n = \frac{1}{\pi} \int_{-\pi}^{\pi} F(x) \cos nx \, dx, \quad B_n = \frac{1}{\pi} \int_{-\pi}^{\pi} F(x) \sin nx \, dx.$$

When $n \neq 0$, we integrate the last two integrals by parts, using the fact that F is continuous and F' is sectionally continuous. Thus

$$A_n = \frac{1}{n\pi} [F(x) \sin nx]_{-\pi}^{\pi} - \frac{1}{n\pi} \int_{-\pi}^{\pi} F'(x) \sin nx \, dx$$

$$= -\frac{1}{n\pi} \int_{-\pi}^{\pi} \left[f(x) - \frac{1}{2} a_0 \right] \sin nx \, dx = -\frac{1}{n} b_n.$$

Similarly, $B_n = a_n/n$; hence

$$(4) \qquad F(x) = \frac{1}{2} A_0 + \sum_{n=1}^{\infty} \frac{1}{n} (a_n \sin nx - b_n \cos nx)$$

when $-\pi \leqq x \leqq \pi$. But since $F(\pi) = \frac{1}{2} a_0 \pi$,

$$\frac{1}{2} a_0 \pi = \frac{1}{2} A_0 - \sum_{n=1}^{\infty} \frac{1}{n} b_n \cos n\pi.$$

With the value of A_0 given here equation (4) becomes

$$F(x) = \frac{a_0 \pi}{2} + \sum_{n=1}^{\infty} \frac{1}{n} [a_n \sin nx - b_n(\cos nx - \cos n\pi)].$$

In view of equation (3), equation (2) follows at once.

The theorem can be written for the integral from x_0 to x, when $-\pi \leqq x_0 \leqq \pi$ and $-\pi \leqq x \leqq \pi$, by noting that

$$\int_{x_0}^{x} f(\xi)\, d\xi = \int_{-\pi}^{x} f(\xi)\, d\xi - \int_{-\pi}^{x_0} f(\xi)\, d\xi.$$

PROBLEMS

1. Note that the function f, where $f(x) = 0$ when $-\pi \leqq x \leqq 0$ and $f(x) = \sin x$ when $0 \leqq x \leqq \pi$, satisfies all the conditions in Theorems 1 and 2. Verify directly from the Fourier series for f, found in Problem 4, Sec. 38, that the series converges uniformly for all x. Also, state why that series is differentiable on the interval $(-\pi,\pi)$ except at the points $x = 0$, $\pm\pi$, and describe the function represented by the derived series for all x.

2. Differentiate the Fourier cosine series for the function $f(x) = x$ on the interval $0 < x < \pi$ (Problem 6, Sec. 38) to obtain the Fourier sine series expansion of the function $f'(x) = 1$ on that interval. State why the procedure is reliable in this case.

3. State Theorem 3 as it applies to Fourier sine series on the interval $(0,\pi)$. Point out, in particular, why the conditions $f(0) = f(\pi) = 0$ are present in this case.

4. Prove that the Fourier coefficients a_n and b_n for the function f in Theorem 1 satisfy the conditions

$$\lim_{n \to \infty} na_n = 0, \qquad \lim_{n \to \infty} nb_n = 0.$$

5. Integrate from 0 to x $(-\pi \leqq x \leqq \pi)$ the Fourier series obtained (a) in Problem 2, Sec. 38; (b) in Problem 3, Sec. 38.   Describe the functions represented by the new series.

49. More General Conditions.   A few of the many more general results in the theory of Fourier series will be noted here. They are stated without proof; our purpose is only to inform the reader of the existence of such theorems.   We first introduce some concepts involved in the theorems.

A function g defined at every point of a closed interval is *monotone nondecreasing* there if its value $g(x)$ never decreases when x is increased.   One example, on the interval $-\pi \leqq x \leqq \pi$, is the step function G, where

$$(1) \quad G(x) = \begin{cases} 0 & \text{when } -\pi \leqq x < \tfrac{1}{2}\pi \text{ and } G(\pi) = \pi, \\[2mm] \dfrac{n}{n+1}\,\pi & \text{when } \dfrac{n}{n+1}\,\pi \leqq x < \dfrac{n+1}{n+2}\,\pi \\[2mm] & \qquad (n = 1, 2, \ldots). \end{cases}$$

This function is not sectionally continuous on the interval; graphically, it has steps up to the line $y = x$ at the infinite set of points $x = n\pi/(n+1)$.   The   function $-G$ is   monotone nonincreasing.

A function f of *bounded variation* on an interval $a \leqq x \leqq b$ can be defined as one that is the sum of two monotone functions g and h,

$$f(x) = g(x) + h(x) \qquad (a \leqq x \leqq b),$$

where g is nondecreasing and h is nonincreasing.   Each such function f has the following properties.[1]   The one-sided limits $f(x+0)$ and $f(x-0)$ from the interior of the interval exist at each point; f has at most a countable infinity of discontinuities on the interval, and f is bounded and integrable over the interval.

a. Another Fourier Theorem.   Let f denote a periodic function of period 2π such that $\int_{-\pi}^{\pi} f(x)\,dx$ exists.   If the integral is improper, let it be absolutely convergent.   Then at each point x which is interior to an interval on which f is of bounded variation,

---

[1] See, for instance, P. Franklin, "Treatise on Advanced Calculus," pp. 255ff., 1940.

the Fourier series for the function converges to the value

$$\tfrac{1}{2}[f(x + 0) + f(x - 0)].$$

The periodic extension of the function G defined on the interval $(-\pi,\pi)$ by conditions (1), for example, satisfies the conditions in this theorem, but not those in our Fourier Theorem 1 of Chap. 4. For that periodic function the series converges at every point to the mean value of the limits from the right and left.

The cosine series for the function $\sqrt{x}$ on the interval $0 \leqq x \leqq \pi$ converges to $\sqrt{x}$ over that interval, including the point $x = 0$ where the right-hand derivative fails to exist. Also, the series for the unbounded function $x^{-\frac{1}{3}}$ on the interval $(-\pi,\pi)$ represents the function over the interval except at the point $x = 0$.

b. Uniform Convergence. Let the periodic integrable function f described in (*a*) above satisfy this additional condition: on some interval $a \leqq x \leqq b$ the function is continuous and of bounded variation. Then its Fourier series converges uniformly to $f(x)$ over each closed interval interior to the interval (*a,b*).

The series for the function $f(x) = x$ on the interval $(-\pi,\pi)$, for example, converges uniformly to x on the interval $-3 \leqq x \leqq 3$.

We noted earlier that the partial sums $S_m(x)$ of the Fourier series for a periodic function f cannot approach $f(x)$ uniformly over an interval that contains a point where f is discontinuous. The nature of the deviation of $S_m(x)$ from $f(x)$ on such intervals is known as the *Gibbs phenomenon.*[1]

c. Integration. The Parseval relation

$$(2) \qquad \frac{1}{2}\, a_0{}^2 + \sum_{n=1}^{\infty} (a_n{}^2 + b_n{}^2) = \frac{1}{\pi} \int_{-\pi}^{\pi} [f(x)]^2\, dx$$

is satisfied whenever f is bounded and integrable over the interval $(-\pi,\pi)$. That is, this series of squares of the Fourier coefficients of f converges to $\pi^{-1}\|f\|^2$.

Proofs of the theorems stated under (*a*), (*b*), and (*c*) above are given in some of the books listed in the Appendix.[2]

---

[1] See H. S. Carslaw, "Theory of Fourier's Series and Integrals."

[2] See, in particular, Whittaker and Watson, "Modern Analysis," chap. 9.

Let a function F also satisfy the conditions stated under (c), and let A_n and B_n denote its Fourier coefficients.   Then $a_n + A_n$ and $b_n + B_n$ are the coefficients for $f + F$ and, according to Parseval's equation,

$$\frac{1}{\pi} \int_{-\pi}^{\pi} [f(x) + F(x)]^2 \, dx = \frac{1}{2} (a_0 + A_0)^2$$

$$+ \sum_{n=1}^{\infty} [(a_n + A_n)^2 + (b_n + B_n)^2].$$

When we subtract the corresponding equation for $f - F$ we obtain *Parseval's equation for the inner product:*

$$(3) \quad \frac{1}{\pi} \int_{-\pi}^{\pi} f(x)F(x) \, dx = \frac{1}{2} a_0 A_0 + \sum_{n=1}^{\infty} (a_n A_n + b_n B_n).$$

In equation (3) let us write

$$F(x) = \begin{cases} g(x) & \text{when } -\pi < x < t \\ 0 & \text{when } t < x < \pi \end{cases} \quad (-\pi \leqq t \leqq \pi),$$

where g is bounded and integrable on the interval $(-\pi,\pi)$.   Then equation (3) takes the form

$$(4) \quad \int_{-\pi}^{t} f(x)g(x) \, dx = \tfrac{1}{2}a_0 \int_{-\pi}^{t} g(x) \, dx$$

$$+ \sum_{n=1}^{\infty} \left[a_n \int_{-\pi}^{t} g(x) \cos nx \, dx + b_n \int_{-\pi}^{t} g(x) \sin nx \, dx \right].$$

So it follows from statement (c) that if the Fourier series corresponding to any bounded integrable function f is multiplied by any other function g of the same class and then integrated term by term, the resulting series converges to the integral of the product $f(x)g(x)$.   When $g(x) = 1$, we have a general theorem for the integration of a Fourier series.

As noted earlier (Sec. 29), statement (c) implies that, in the space of bounded integrable functions on the interval $(-\pi,\pi)$, the orthonormal set of trigonometric functions (4), Sec. 35, is closed in the sense of mean convergence.

CHAPTER 6

FOURIER INTEGRALS

50. The Fourier Integral Formula. We have shown that the Fourier series

$$(1) \qquad \frac{1}{2c} \int_{-c}^{c} f(\xi) \, d\xi + \frac{1}{c} \sum_{n=1}^{\infty} \int_{-c}^{c} f(\xi) \cos \left[\frac{n\pi}{c} (\xi - x) \right] d\xi$$

converges to $f(x)$ when $-c < x < c$, provided the function f satisfies certain conditions on the interval $(-c,c)$. It is sufficient that f be sectionally continuous on the interval and, at each interior point x, that it has one-sided derivatives and is defined as the mean of its limits $f(x + 0)$ and $f(x - 0)$.

Suppose f satisfies such conditions on every finite interval. Then c may be given any fixed value, arbitrarily large but finite, and series (1) will represent $f(x)$ over the large interval $-c < x < c$. But that series representation cannot apply over the rest of the x axis unless f is periodic with period $2c$ because the sum of the series has that periodicity.

To *indicate* a representation that may be valid for all real x when f is not periodic, it is natural to try to extend the above representation by letting c tend to infinity. The first term in series (1) would then vanish, provided that f is such that the integral $\int_{-\infty}^{\infty} f(\xi) \, d\xi$ exists. We write $\Delta\alpha = \pi/c$; then the remaining terms take the form

$$(2) \qquad \frac{1}{\pi} \sum_{n=1}^{\infty} \Delta\alpha \int_{-c}^{c} f(\xi) \cos [n\Delta\alpha(\xi - x)] \, d\xi \qquad \left(c = \frac{\pi}{\Delta\alpha} \right).$$

In terms of the function g_c, where

$$(3) \qquad g_c(\alpha,x) = \int_{-c}^{c} f(\xi) \cos [\alpha(\xi - x)] \, d\xi \qquad \left(c = \frac{\pi}{\Delta\alpha} \right),$$

the series in expression (2) becomes

$$(4) \qquad \sum_{n=1}^{\infty} g_c(n\Delta\alpha, x)\, \Delta\alpha.$$

Let the value of x be fixed. When $\Delta\alpha$ is a small positive number, the points $n\Delta\alpha$ are equally spaced along the entire positive α axis, so the sum of the series (4) may approximate the integral

$$(5) \qquad \int_0^{\infty} g_c(\alpha, x)\, d\alpha,$$

where c is large, or the integral $\int_0^{\infty} g_{\infty}(\alpha, x)\, d\alpha$. Note, however, that the limit of series (4), as $\Delta\alpha \to 0$, is *not* the definition of the improper integral (5) even if c could be kept fixed; moreover, the function g_c changes with $\Delta\alpha$ because $c = \pi/\Delta\alpha$.

The above manipulations merely *suggest* that under appropriate conditions on f the function may have the representation

$$(6) \qquad f(x) = \frac{1}{\pi} \int_0^{\infty} \int_{-\infty}^{\infty} f(\xi) \cos\left[\alpha(\xi - x)\right] d\xi\, d\alpha$$

$$(-\infty < x < \infty).$$

This is the *Fourier integral formula* for the function f, to be established in the following section.

The formula expresses $f(x)$ in terms of sine and cosine functions of x, in the form

$$(7) \qquad f(x) = \int_0^{\infty} \left[A(\alpha) \cos \alpha x + B(\alpha) \sin \alpha x\right] d\alpha$$

$$(-\infty < x < \infty),$$

where

$$(8) \qquad \begin{aligned} A(\alpha) &= \frac{1}{\pi} \int_{-\infty}^{\infty} f(\xi) \cos \alpha\xi\, d\xi, \\ B(\alpha) &= \frac{1}{\pi} \int_{-\infty}^{\infty} f(\xi) \sin \alpha\xi\, d\xi. \end{aligned}$$

Formulas (7) and (8) bear a resemblance to Fourier series representations and formulas for the coefficients a_n and b_n.

51. A Fourier Integral Theorem. The following theorem gives conditions on f under which the Fourier integral formula is valid.

Theorem 1.   *Let f denote a function that is sectionally continuous on every finite interval on the x axis and defined as $\frac{1}{2}[f(x_0 + 0) + f(x_0 - 0)]$ at each point of discontinuity x_0; furthermore, let f be such that*

$$\int_{-\infty}^{\infty} |f(x)| \, dx$$

exists.   Then   at   every   point x where   its   one-sided   derivatives $f'_R(x)$ and $f'_L(x)$ exist, the function is represented by the Fourier integral formula

$$(1) \qquad f(x) = \frac{1}{\pi} \int_0^{\infty} \int_{-\infty}^{\infty} f(\xi) \cos [\alpha(\xi - x)] \, d\xi \, d\alpha$$

$$(-\infty < x < \infty).$$

Under the conditions stated here we first prove that

$$(2) \qquad \lim_{\beta \to \infty} \int_{-\infty}^{\infty} f(\xi) \frac{\sin [\beta(\xi - x)]}{\xi - x} \, d\xi = \pi f(x),$$

where x is a point at which the one-sided derivatives of f exist. Thus we are generalizing the result in Lemma 3, Sec. 40, so that it applies to improper integrals.

The function $S(t) = t^{-1} \sin t \quad (t \neq 0)$, $S(0) = 1$ is continuous, and $|S(t)| \leq 1$, for all real t (Sec. 39).   It follows readily that the integrand in equation (2),

$$(3) \qquad g(\xi,\beta,x) = \beta f(\xi) \frac{\sin [\beta(\xi - x)]}{\beta(\xi - x)}$$

is a sectionally continuous function of ξ on each interval and that $|g(\xi,\beta,x)| \leq |\beta f(\xi)|$.   Since f is absolutely integrable over the ξ axis, the improper integral in equation (2) exists.   We write

$$\int_{-\infty}^{\infty} g(\xi,\beta,x) \, d\xi - \pi f(x) = h(\beta,x)$$

and establish the limit (2) by showing that, for the fixed value of x, to each positive number ϵ there corresponds a number β_ϵ such that

$$(4) \qquad\qquad |h(\beta,x)| < \epsilon \qquad\qquad \text{whenever } \beta > \beta_\epsilon.$$

If a and b denote numbers such that the interval $a < \xi < b$

contains the point $\xi = x$, then

$$|h(\beta,x)| = \left| \int_{-\infty}^{a} g\, d\xi + \int_{a}^{b} g\, d\xi - \pi f(x) + \int_{b}^{\infty} g\, d\xi \right|$$

$$\leqq \int_{-\infty}^{a} |g|\, d\xi + \left| \int_{a}^{b} g\, d\xi - \pi f(x) \right| + \int_{b}^{\infty} |g|\, d\xi$$

and $\displaystyle \int_{-\infty}^{a} |g|\, d\xi \leqq \int_{-\infty}^{a} \frac{|f(\xi)|}{|x - \xi|}\, d\xi \leqq \frac{1}{x - a} \int_{-\infty}^{a} |f(\xi)|\, d\xi.$

The last member is independent of β and tends to zero as $a \to -\infty$. Hence for the given number ϵ a value a_ϵ of a, independent of β, exists such that

$$\int_{-\infty}^{a_\epsilon} |g(\xi,\beta,x)|\, d\xi < \frac{\epsilon}{3}.$$

Likewise a number b_ϵ, independent of β, exists such that

$$\int_{b_\epsilon}^{\infty} |g(\xi,\beta,x)|\, d\xi < \frac{\epsilon}{3},$$

and consequently

(5) $$|h(\beta,x)| < \tfrac{2}{3}\epsilon + \left| \int_{a_\epsilon}^{b_\epsilon} g\, d\xi - \pi f(x) \right|.$$

According to Lemma 3, Sec. 40, there is a number β_ϵ such that the final absolute value in condition (5) is less than $\epsilon/3$ when $\beta > \beta_\epsilon$. This establishes condition (4) and the limit (2).

We can write the fraction in the integrand of formula (2) as an integral and present that formula in the form

(6) $$\pi f(x) = \lim_{\beta \to \infty} \int_{-\infty}^{\infty} f(\xi) \int_{0}^{\beta} \cos\,[\alpha(\xi - x)]\, d\alpha\, d\xi.$$

For all α the integral

$$\int_{-\infty}^{\infty} f(\xi) \cos\,[\alpha(\xi - x)]\, d\xi$$

is uniformly convergent with respect to α according to the Weierstrass test. Thus

$$\int_{0}^{\beta} \int_{-\infty}^{\infty} f(\xi) \cos\,[\alpha(\xi - x)]\, d\xi\, d\alpha$$
$$= \int_{-\infty}^{\infty} f(\xi) \int_{0}^{\beta} \cos\,[\alpha(\xi - x)]\, d\alpha\, d\xi$$

so that formula (6) can be written in the form

$$\pi f(x) = \lim_{\beta \to \infty} \int_0^\beta \int_{-\infty}^\infty f(\xi) \cos [\alpha(\xi - x)] \, d\xi \, d\alpha,$$

which is the Fourier integral formula (1).

Other conditions of validity of the formula are known.[1] The conditions on f can be relaxed when Lebesgue integrals are used in place of our Riemann integrals.

Under the conditions stated in Theorem 1 the integrals in formulas (8), Sec. 50, for the coefficients $A(\alpha)$ and $B(\alpha)$, exist; hence the integral formula can be written in the form (7) of Sec. 50.

PROBLEMS

1. Verify that all conditions in Theorem 1 are satisfied by this function f:

$$\begin{aligned}
f(x) &= 1 & &\text{when } |x| < 1, \\
f(x) &= 0 & &\text{when } |x| > 1, \\
f(1) &= f(-1) = \tfrac{1}{2};
\end{aligned}$$

hence show that for every x $(-\infty < x < \infty)$ it is true that

$$f(x) = \frac{1}{\pi} \int_0^\infty \frac{\sin [\alpha(1 - x)] + \sin [\alpha(1 + x)]}{\alpha} \, d\alpha = \frac{2}{\pi} \int_0^\infty \frac{\sin \alpha \cos \alpha x}{\alpha} \, d\alpha.$$

2. Use the integration formula derived in Sec. 39 to show that the integral $\int_0^\infty t^{-1} \sin kt \, dt$ has the value $\tfrac{1}{2}\pi$ when $k > 0$, $-\tfrac{1}{2}\pi$ when $k < 0$, and zero when $k = 0$. Thus evaluate the first integral written in Problem 1 when $|x| < 1$, when $|x| > 1$, and when $|x| = 1$ to make a direct verification of the Fourier integral representation found there for the step function f.

3. If $f(x) = 0$ when $x < 0$ and $f(x) = e^{-x}$ when $x > 0$, and $f(0) = \tfrac{1}{2}$, show that f satisfies all conditions in our Fourier integral theorem and therefore, for each value of x, that

$$f(x) = \frac{1}{\pi} \int_0^\infty \frac{\cos \alpha x + \alpha \sin \alpha x}{1 + \alpha^2} \, d\alpha \quad (-\infty < x < \infty).$$

Verify this representation directly at the point $x = 0$.

[1] See, for instance, Carslaw, "Theory of Fourier's Series and Integrals," pp. 315ff., and Titchmarsh, "Theory of Fourier Integrals."

4. Prove that

$$\exp\left(-|x|\right) = \frac{2}{\pi}\int_0^\infty \frac{\cos \alpha x}{1 + \alpha^2}\, d\alpha \qquad (-\infty < x < \infty).$$

5. If $f(x) = 0$ when $x \leq 0$ and when $x \geq \pi$, and $f(x) = \sin x$ when $0 \leq x \leq \pi$, prove that

$$f(x) = \frac{1}{\pi}\int_0^\infty \frac{\cos \alpha x + \cos\left[\alpha(\pi - x)\right]}{1 - \alpha^2}\, d\alpha \qquad (-\infty < x < \infty).$$

In particular, write $x = \frac{1}{2}\pi$ to show that

$$\int_0^\infty \frac{\cos \frac{1}{2}\pi\alpha}{1 - \alpha^2}\, d\alpha = \frac{\pi}{2}.$$

6. Show why the Fourier integral formula fails to represent the function f if $f(x) = 1$ $(-\infty < x < \infty)$. Also, note which condition in Theorem 1 is not satisfied by that function.

7. In the Fourier integral formula (1), Sec. 51, show why it is not possible to interchange the order of integration with respect to ξ and α, except by writing the form (6) of the formula.

8. If a nonzero function f is periodic for all x, point out why $f(x)$ and $|f(x)|$ are not integrable from $x = -\infty$ to $x = \infty$.

52. Sine or Cosine Forms. Let f be an *odd* function that satisfies the conditions of Theorem 1. Then $f(\xi) \cos \alpha\xi$ and $f(\xi) \sin \alpha\xi$ are odd and even functions of ξ, respectively, whose improper integrals over the entire ξ axis exist. Therefore

$$A(\alpha) = \frac{1}{\pi}\int_{-\infty}^\infty f(\xi) \cos \alpha\xi\, d\xi = 0,$$

$$B(\alpha) = \frac{1}{\pi}\int_{-\infty}^\infty f(\xi) \sin \alpha\xi\, d\xi = \frac{2}{\pi}\int_0^\infty f(\xi) \sin \alpha\xi\, d\xi,$$

and the Fourier integral formula

$$(1) \quad f(x) = \int_0^\infty \left[A(\alpha) \cos \alpha x + B(\alpha) \sin \alpha x\right] d\alpha$$

$$(-\infty < x < \infty)$$

reduces to the *Fourier sine integral formula*

$$(2) \qquad f(x) = \frac{2}{\pi}\int_0^\infty \sin \alpha x \int_0^\infty f(\xi) \sin \alpha\xi\, d\xi\, d\alpha.$$

If, instead, f is an *even* function, then

$$A(\alpha) = \frac{2}{\pi}\int_0^\infty f(\xi) \cos \alpha\xi\, d\xi, \qquad B(\alpha) = 0$$

and formula (1) becomes the *Fourier cosine integral formula*

$$(3) \qquad f(x) = \frac{2}{\pi} \int_0^\infty \cos \alpha x \int_0^\infty f(\xi) \cos \alpha \xi \, d\xi \, d\alpha.$$

In case $f(x)$ is defined only on the semi-infinite interval $x > 0$, formulas (2) and (3) apply to the odd and even extensions of f, respectively, to represent $f(x)$ when $x > 0$ under the following conditions.

Corollary 1. *Let a function f be sectionally continuous on every finite interval on the positive x axis and defined as the mean of its limits $f(x_0 + 0)$ and $f(x_0 - 0)$ at each point x_0 $(x_0 > 0)$ where f is discontinuous. Suppose further that $\int_0^\infty |f(x)| \, dx$ exists. Then at every point x $(x > 0)$ where its right- and left-hand derivatives exist, the function is represented by its Fourier sine integral formula (2) and by its Fourier cosine integral formula (3).*

Under those conditions, when $x = 0$, the cosine integral formula (3) represents $f(+0)$ in case $f'_R(0)$ exists, as we see by considering the even extension of f. The right-hand member of the sine integral formula (2) clearly has the value zero when $x = 0$.

The eigenvalue problem

$$(4) \qquad \begin{aligned} X''(x) + \lambda X(x) &= 0 & (x > 0); \\ X(0) = 0, \quad |X(x)| &< M & (x > 0), \end{aligned}$$

where M is some positive constant, is *singular* because its fundamental interval $x > 0$ is unbounded.   When $\lambda \neq 0$, a solution of the differential equation satisfying the condition $X(0) = 0$ is $\sin x \sqrt{\lambda}$, and this is bounded for all positive x if and only if $\sqrt{\lambda}$ has real values.   We write $\sqrt{\lambda} = \alpha$; then, except for a constant factor, the eigenfunctions are $X = \sin \alpha x$, where α takes on all positive real values.   The eigenvalues $\lambda = \alpha^2$ are continuous rather than discrete.   Although the eigenfunctions $\sin \alpha x$ have no orthogonality property, the Fourier sine integral formula (2) gives a representation of functions f on the interval $x > 0$ as a generalized linear combination of those eigenfunctions.

Likewise, $\lambda = \alpha^2$ $(\alpha \geqq 0)$ and $X = \cos \alpha x$ are the eigenvalues and eigenfunctions of the singular problem

$$(5) \qquad \begin{aligned} X''(x) + \lambda X(x) &= 0 & (x > 0); \\ X'(0) = 0, \quad |X(x)| &< M & (x > 0), \end{aligned}$$

and formula (3) represents functions f in terms of $\cos \alpha x$.

53. The Exponential Form.   Under the conditions stated in
Theorem 1 the Fourier integral formula can be written

$$(1) \qquad f(x) = \frac{1}{2\pi} \lim_{\beta \to \infty} \int_0^\beta \int_{-\infty}^\infty 2f(\xi) \cos [\alpha(\xi - x)] \, d\xi \, d\alpha.$$

Let the cosine function be expressed in terms of exponential
functions.   Since f is absolutely integrable, the integrals

$$\int_{-\infty}^\infty f(\xi)e^{i\alpha\xi} \, d\xi, \qquad \int_{-\infty}^\infty f(\xi)e^{-i\alpha\xi} \, d\xi$$

converge uniformly with respect to α for all positive α, according
to the Weierstrass test.   That uniform convergence together with
the sectional continuity of f implies that those integrals represent
continuous functions of α (see Problem 9, Sec. 54).   Thus the
iterated integral in formula (1) can be written as the sum

$$(2) \qquad \int_0^\beta e^{-i\alpha x} \int_{-\infty}^\infty f(\xi)e^{i\alpha\xi} \, d\xi \, d\alpha + \int_0^\beta e^{i\alpha x} \int_{-\infty}^\infty f(\xi)e^{-i\alpha\xi} \, d\xi \, d\alpha$$

since the definite integrals with respect to α exist as integrals
of continuous functions of α.   (The improper integrals from
$\alpha = 0$ to $\alpha = \infty$ may not exist, however.)

With the substitution $\alpha' = -\alpha$ the second term of the sum (2)
takes the form

$$\int_{-\beta}^0 e^{-i\alpha' x} \int_{-\infty}^\infty f(\xi)e^{i\alpha'\xi} \, d\xi \, d\alpha'.$$

After we drop the primes here, the sum (2) can be written

$$\int_{-\beta}^\beta e^{-i\alpha x} \int_{-\infty}^\infty f(\xi)e^{i\alpha\xi} \, d\xi \, d\alpha.$$

Thus the *exponential form of the Fourier integral formula* (1) is

$$(3) \qquad f(x) = \frac{1}{2\pi} \lim_{\beta \to \infty} \int_{-\beta}^\beta e^{-i\alpha x} \int_{-\infty}^\infty f(\xi)e^{i\alpha\xi} \, d\xi \, d\alpha,$$

where $-\infty < x < \infty$.   The minus sign in the exponent $-i\alpha x$
can be shifted to the exponent $i\alpha\xi$ by using $-\alpha$ as a variable of
integration.

The limit in formula (3) is called the *Cauchy principal value*
of the improper integral from $-\infty$ to ∞ with respect to α.   The
principal value may exist when the improper integral does not.
The integral $\int_{-\infty}^\infty \alpha \, d\alpha$, for example, has no value; but its princi-

pal value does exist, and it is equal to zero because $\int_{-\beta}^{\beta} \alpha \, d\alpha = 0$ for each value of β. In case the improper integral does exist it is equal to its principal value.

Unless further conditions are imposed on f, the principal value used in formula (3) cannot be replaced by the improper integral itself, as an example will show. The function

$$(4) \qquad \begin{aligned} f(x) &= 0 && \text{when } x < 0, \\ f(x) &= e^{-x} && \text{when } x > 0, \\ f(0) &= \tfrac{1}{2} \end{aligned}$$

satisfies all conditions in our Fourier integral theorem. It is represented by formula (3) at every point x, in particular, at $x = 0$. In this case

$$(5) \qquad \int_{-\infty}^{\infty} f(\xi) e^{i\alpha\xi} \, d\xi = \int_{0}^{\infty} e^{-\xi(1-i\alpha)} \, d\xi = \frac{1}{1 - i\alpha} = \frac{1 + i\alpha}{1 + \alpha^2}.$$

Thus when $x = 0$ in formula (3), the iterated integral becomes

$$\int_{-\beta}^{\beta} \frac{1 + i\alpha}{1 + \alpha^2} \, d\alpha = \left[\arctan \alpha + \frac{i}{2} \log (1 + \alpha^2) \right]_{-\beta}^{\beta} = 2 \arctan \beta,$$

which has the limit π as $\beta \to \infty$. The right-hand member of formula (3) therefore has the value $\tfrac{1}{2}$, which is $f(0)$. But the improper integral of the function (5), from $-\infty$ to ∞, does not exist since the function $\alpha(1 + \alpha^2)^{-1}$ is not integrable from 0 to ∞.

The functions $e^{-i\alpha x}$ $(-\infty < \alpha < \infty)$ are eigenfunctions of the singular eigenvalue problem

$$(6) \qquad X''(x) + \lambda X(x) = 0, \qquad |X(x)| < M \quad (-\infty < x < \infty),$$

where M is some constant. The eigenvalues $\lambda = \alpha^2$ consist of all real nonnegative numbers, and the representation of functions f in terms of the eigenfunctions is given by formula (3).

54. Fourier Transforms. The Fourier sine integral formula can be written in the form

$$(1) \qquad f(x) = \frac{2}{\pi} \int_{0}^{\infty} F_s(\alpha) \sin \alpha x \, d\alpha \qquad (x > 0),$$

where

$$(2) \qquad F_s(\alpha) = \int_{0}^{\infty} f(\xi) \sin \alpha \xi \, d\xi \qquad (\alpha > 0).$$

If f is a given function, then equation (1) is an *integral equation* in the function F_s, an equation containing the unknown function in the integrand.   It is a *singular* integral equation because the integral is improper.   Equation (2) gives a solution of that integral equation when f satisfies the conditions stated in Corollary 1, as we see by substituting the expression (2) for F_s into equation (1).

The function F_s defined by equation (2) is the *Fourier sine transform* of the function f.   The transformation (2), which we may abbreviate as

$$(3) \qquad\qquad F_s(\alpha) = S_\alpha\{f\},$$

sets up a correspondence between functions f and F_s.   Functions f satisfying the conditions of Corollary 1 have transforms F_s such that formula (1) gives f in terms of its transform; that is, formula (1) gives the inverse transformation.

Suppose that f, f', and f'' are continuous when $x \geqq 0$, that $f(x)$ and $f'(x)$ tend to zero as $x \to \infty$, and that f is absolutely integrable from 0 to ∞.   Then successive integrations by parts show that (Problem 7)

$$(4) \qquad\qquad S_\alpha\{f''(x)\} = -\alpha^2 F_s(\alpha) + \alpha f(0).$$

This is the basic operational property of the sine transformation, that of replacing the differential form $f''(x)$ by an algebraic form in the transform F_s, α, and the initial value $f(0)$.

Property (4), together with other operational properties of the transformation, lead to the reduction of certain types of boundary value problems to simpler problems in the transforms of the unknown functions.   The development of this operational method, using the various Fourier transformations and the related Laplace transformation, is treated elsewhere.[1]

The *Fourier cosine transform* F_c of a function f is

$$(5) \qquad\qquad F_c(\alpha) = \int_0^\infty f(\xi) \cos \alpha\xi \, d\xi = C_\alpha\{f\} \qquad (\alpha > 0).$$

The inverse of the transformation $C_\alpha\{f\}$ is given by the Fourier cosine integral formula,

$$(6) \qquad\qquad f(x) = \frac{2}{\pi} \int_0^\infty F_c(\alpha) \cos \alpha x \, d\alpha \qquad (x > 0).$$

The basic operational property, valid under the conditions given

---

[1] Churchill, R. V., "Operational Mathematics," 2d ed., 1958.

above for property (4), involves $f'(0)$:

(7) $$C_\alpha\{f''(x)\} = -\alpha^2 F_c(\alpha) - f'(0).$$

Other Fourier transformations include the exponential form

(8) $$\int_{-\infty}^{\infty} f(\xi)e^{i\alpha\xi}\,d\xi = F(\alpha) \qquad (-\infty < \alpha < \infty)$$

whose inverse is given by formula (3), Sec. 53, and also certain integral transformations over bounded intervals called finite Fourier transformations.

PROBLEMS

1. If $f(x) = 1$ when $0 < x < k$ and $f(x) = 0$ when $x > k$, and $f(k) = \frac{1}{2}$, show that the Fourier sine integral formula applies to f to give the representation

$$f(x) = \frac{2}{\pi}\int_0^\infty \frac{1 - \cos k\alpha}{\alpha}\sin \alpha x\,d\alpha \qquad (x > 0).$$

2. Use Corollary 1 and the Fourier sine integral formula to show that

$$e^{-x}\cos x = \frac{2}{\pi}\int_0^\infty \frac{\alpha^3\sin \alpha x}{\alpha^4 + 4}\,d\alpha \qquad (x > 0).$$

3. Use the Fourier cosine integral formula and prove that

$$e^{-x}\cos x = \frac{2}{\pi}\int_0^\infty \frac{\alpha^2 + 2}{\alpha^4 + 4}\cos \alpha x\,d\alpha \qquad (x \geqq 0).$$

4. Apply the operational property (4), Sec. 54, to the function e^{-kx}, where k is a positive constant, to show that the sine transform of that function is $\alpha(\alpha^2 + k^2)^{-1}$; hence obtain the representation

$$e^{-kx} = \frac{2}{\pi}\int_0^\infty \frac{\alpha\sin \alpha x}{\alpha^2 + k^2}\,d\alpha \qquad (x > 0, k > 0).$$

5. Use the operational property (7), Sec. 54, to prove that the cosine transform of e^{-kx}, where k is a positive constant, is $k(\alpha^2 + k^2)^{-1}$; hence show that

$$e^{-kx} = \frac{2k}{\pi}\int_0^\infty \frac{\cos \alpha x}{\alpha^2 + k^2}\,d\alpha \qquad (x \geqq 0, k > 0).$$

6. Establish the representation

$$e^{-x} - e^{-2x} = \frac{6}{\pi}\int_0^\infty \frac{\alpha\sin \alpha x}{(\alpha^2 + 1)(\alpha^2 + 4)}\,d\alpha \qquad (x \geqq 0).$$

In the boundary value problem treated in Sec. 15, show that the function

$$u(x,y) = \int_0^\infty g(\alpha)e^{-\alpha y} \sin \alpha x \, d\alpha \qquad (x \geq 0, \, y \geq 0)$$

satisfies the additional boundary condition

$$u(x,0) = e^{-x} - e^{-2x} \qquad\qquad (x \geq 0)$$

if g is the absolutely integrable function

$$g(\alpha) = \frac{6}{\pi} \frac{\alpha}{(\alpha^2 + 1)(\alpha^2 + 4)} \qquad\qquad (\alpha \geq 0).$$

7. When f, f', and f'' are continuous ($x \geq 0$), use integration by parts to prove that, for each positive constant c,

$$\int_0^c f''(x) \sin \alpha x \, dx = f'(c) \sin \alpha c - \alpha f(c) \cos \alpha c + \alpha f(0)$$
$$- \alpha^2 \int_0^c f(x) \sin \alpha x \, dx.$$

[The continuity of f'' here can be replaced by the condition that f'' be sectionally continuous on each interval $(0,c)$.] Assuming also that $f(x)$ and $f'(x)$ tend to zero as $x \to \infty$ and that the Fourier sine transform $F_s(\alpha)$ of f exists, show that the right-hand member of the above equation has the limit $\alpha f(0) - \alpha^2 F_s(\alpha)$ as $c \to \infty$. Deduce that the sine transform of f'' exists and satisfies the operational property (4), Sec. 54.

8. Derive the operational property (7), Sec. 54, for the cosine transform of f'' (compare Problem 7).

9. When $f(\xi)$ is sectionally continuous on each interval $(-c,c)$ of the ξ axis, state why the function

$$\phi(\alpha) = \int_{-c}^c f(\xi)e^{i\alpha\xi} \, d\xi$$

is continuous for all real α. When, in addition, f is absolutely integrable along the entire ξ axis, then the integral

$$\int_{-\infty}^\infty f(\xi)e^{i\alpha\xi} \, d\xi = F(\alpha)$$

is uniformly convergent with respect to α. Write

$$F(\alpha) = \phi(\alpha) + \int_{-\infty}^{-c} f(\xi)e^{i\alpha\xi} \, d\xi + \int_c^\infty f(\xi)e^{i\alpha\xi} \, d\xi$$

and prove that $F(\alpha + \Delta\alpha) - F(\alpha) \to 0$ as $\Delta\alpha \to 0$ by first making c large, independent of α and $\Delta\alpha$, and then taking $|\Delta\alpha|$ small to make $|\Delta\phi|$ small, thus establishing the continuity of $F(\alpha)$ used in Sec. 53.

10. Establish the following operational property of the Fourier transformation

$$E_\alpha\{f\} = \int_{-\infty}^{\infty} f(x)e^{i\alpha x}\,dx = F(\alpha) \quad (-\infty < \alpha < \infty).$$

If f, f', and f'' are everywhere continuous, if the transform $F(\alpha)$ of f exists, and if $f(x)$ and $f'(x)$ both tend to zero as $|x| \to \infty$, then

$$E_\alpha\{f''(x)\} = -\alpha^2 F(\alpha) \qquad (-\infty < \alpha < \infty).$$

11. Verify the Fourier sine integral representation

$$\frac{x}{x^2 + k^2} = \frac{2}{\pi} \int_0^\infty \sin \alpha x \int_0^\infty \frac{\xi \sin \alpha \xi}{\xi^2 + k^2}\, d\xi\, d\alpha \qquad (k > 0)$$

by observing that, according to Problem 4, the inner integral here has the value $(\pi/2)e^{-k\alpha}$ and the outer integral is then the sine transform of that exponential function of α. Note that the function $x/(x^2 + k^2)$ is *not* absolutely integrable from $x = 0$ to $x = \infty$.

CHAPTER 7

BOUNDARY VALUE PROBLEMS

55. Formal and Rigorous Solutions. The foregoing theory of representing prescribed functions by Fourier series or integrals enables us to use the method of separation of variables and super-position to solve important types of boundary value problems in partial differential equations. The method is illustrated in this chapter for a variety of boundary value problems which are mathematical formulations of problems in physics.

We illustrate ways of proving that the function we find truly satisfies the partial differential equation and all boundary conditions and continuity requirements. When that is done, our function is rigorously established as *a* solution of the problem. The physical problem may indicate that there should be only *one* solution. In Chap. 10 we shall give some attention to that question of uniqueness of solutions.

Even for some of the simpler problems, the full treatment, consisting of establishing a solution and proving that it is the only solution, may be lengthy or even difficult. We shall make that full treatment in only a few cases. Most of the problems will be solved *formally* in the sense that we may not verify our solution fully or that we may not prove that our solution is unique.

56. The Vibrating String, Initially Displaced. In Sec. 16 we attempted to find a formula for the transverse displacements $y(x,t)$ of a string stretched between the fixed points $(0,0)$ and $(c,0)$, after the string is released at rest from a position $y = f(x)$ in the xy plane (Fig. 11). The function y is therefore required to satisfy all conditions of the boundary value problem

$$(1) \qquad y_{tt}(x,t) = a^2 y_{xx}(x,t) \qquad (0 < x < c, t > 0),$$
$$(2) \qquad y(0,t) = 0, \qquad y(c,t) = 0 \qquad (t \geqq 0),$$
$$(3) \qquad y(x,0) = f(x), \qquad y_t(x,0) = 0 \qquad (0 \leqq x \leqq c).$$

126

We used separation of variables, superposition, and the orthogonality of the functions $\sin(n\pi x/c)$ to arrive at a formal solution

$$(4) \qquad y(x,t) = \sum_{n=1}^{\infty} b_n \sin \frac{n\pi x}{c} \cos \frac{n\pi at}{c},$$

where

$$(5) \qquad b_n = \frac{2}{c} \int_0^c f(x) \sin \frac{n\pi x}{c} \, dx \qquad (n = 1, 2, \ldots).$$

·The given function f is to be continuous on the interval $(0 \le x \le c)$; also, $f(0) = f(c) = 0$. We assume that f' is at least sectionally continuous on the interval. Under those conditions we now know that f is represented by its Fourier sine series on that interval. The coefficients in that series are the numbers b_n given by equation (5). Hence when $t = 0$, the series in

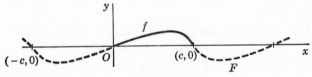

FIG. 11

formula (4) does converge to $f(x)$; that is, $y(x,0) = f(x)$ when $0 \le x \le c$.

The nature of the problem calls for a solution $y(x,t)$ that is continuous in x and t when $0 \le x \le c$ and $t \ge 0$ and such that $y_t(x,t)$ is continuous in t at $t = 0$. Then the prescribed boundary values are limiting values at the boundaries $y(0,t) = y(+0,t)$, $y(c,t) = y(c - 0, t)$, etc.

To show that formula (4) represents a solution we should prove that the series there converges to a continuous function $y(x,t)$ which, together with its partial derivatives, satisfies the wave equation (1) and the boundary conditions. But the series may not be twice differentiable with respect to x and t even though the series has a sum $y(x,t)$ that may satisfy the wave equation. In the case of the plucked string (Sec. 18), for instance, the coefficients b_n are proportional to $n^{-2} \sin(n\pi/2)$ so that, after differentiating the series for $y(x,t)$ twice with respect to either x or t, the resulting series fails to converge.

It is possible to sum the series in formula (4), that is, to represent its sum without using infinite series. This will simplify the verification of the solution.

Since

$$2 \sin \frac{n\pi x}{c} \cos \frac{n\pi a t}{c} = \sin \frac{n\pi(x - at)}{c} + \sin \frac{n\pi(x + at)}{c},$$

equation (4) can be written

$$(6) \quad y = \frac{1}{2} \sum_{n=1}^{\infty} b_n \sin \left[\frac{n\pi}{c} (x - at) \right] + \frac{1}{2} \sum_{n=1}^{\infty} b_n \sin \left[\frac{n\pi}{c} (x + at) \right].$$

Let $F(x)$ be defined for *all* real x by the sine series for f:

$$(7) \qquad\qquad F(x) = \sum_{n=1}^{\infty} b_n \sin \frac{n\pi x}{c} \qquad (-\infty < x < \infty).$$

Then $F(x)$ is the odd periodic extension, with period $2c$, of $f(x)$; that is,

$$(8) \qquad \begin{aligned} F(x) &= f(x) && \text{when } 0 \leqq x \leqq c, \\ F(-x) &= -F(x), & F(x + 2c) &= F(x), && \text{for all } x. \end{aligned}$$

In view of equation (7), formula (6) can be written

$$(9) \qquad\qquad y(x,t) = \tfrac{1}{2}[F(x - at) + F(x + at)].$$

Thus, series (4) is summed with the aid of the function F defined by equations (8). The convergence of series (6) and (4) follows from the convergence of series (7).

Solution Established. Our conditions on f are such that its extension F is continuous for all x (Fig. 11). Hence $F(x - at)$ and $F(x + at)$, and therefore the function y given by formula (9), are continuous functions of x and t for all x and t. From either formula (4) or (9) we find that $y(0,t) = 0$, $y(c,t) = 0$, and $y(x,0) = f(x)$. Note that when $x = c$ in formula (9), we can write $F(c - at) = -F(at - c) = -F(at + c)$; thus $y(c,t) = 0$.

Since $F(-x) = -F(x)$, then $-F'(-x) = -F'(x)$ whenever $F'(x)$ exists, where the prime denotes the derivative with respect to the argument of F. That is, F' is an even function. Likewise F'' is an odd function.

In case f' and f'' are continuous when $0 \leq x \leq c$ and

$$f''(0) = f''(c) = 0,$$

then $F'(x)$ and $F''(x)$ are continuous for all x, as indicated in Fig. 12. Thus

$$y_t(x,t) = \frac{a}{2}[- F'(x - at) + F'(x + at)],$$

y_t is continuous for all x and t, and $y_t(x,0) = 0$. As noted in Sec. 20, $F(x - at)$ and $F(x + at)$ satisfy the wave equation (1); therefore y satisfies that equation, as well as all boundary conditions. The function y given by formula (9) is then established as a solution of our boundary value problem. In Chap. 10 we shall show why it is the only possible solution which, together with its derivatives of first and second order, is everywhere continuous.

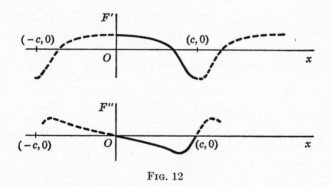

Fig. 12

If the conditions on f' and f'' are relaxed by merely requiring those two functions to be *sectionally continuous*, we find that at each instant t there may be at most a finite number of points x $(0 \leq x \leq c)$ where the partial derivatives of y fail to exist. Except at those points, our function satisfies the wave equation and the condition $y_t(x,0) = 0$. The other boundary conditions are satisfied as before.   In this case we have a solution of our boundary value problem in a broader sense.

57. Discussion of the Solution. From either formula (4) above or its alternate form (9) we can see that for each fixed x the displacement $y(x,t)$ is a periodic function of time t, with the period

$$(1) \qquad\qquad T_0 = \frac{2c}{a}.$$

The period is independent of the initial displacement $f(x)$. Since $a^2 = H/\delta$, where H is the horizontal component of the tension and δ is the mass per unit length, the period varies directly with the length c and with $\sqrt{\delta}$, and inversely with $\sqrt{H}$.

From either formula (4) or (9) it is also evident that, for a given length c and initial displacement $f(x)$, the displacement y depends only on the value of x and the value of the product at. That is, $y = \phi(x,at)$, where the function ϕ is the same function regardless of the value of the constant a. Let a_1 and a_2 denote different values of that constant, and $y_1(x,t)$ and $y_2(x,t)$ the corresponding displacements. Then

$$(2) \qquad y_1(x,t_1) = y_2(x,t_2) \qquad \text{if } a_1t_1 = a_2t_2 \quad (0 \le x \le c).$$

In particular, suppose that only the tensions differ, with values H_1 and H_2. The same set of instantaneous positions is then taken by the string when $H = H_1$ and when $H = H_2$, but the times t_1 and t_2 required to reach any one position have the ratio

$$(3) \qquad\qquad \frac{t_1}{t_2} = \left(\frac{H_2}{H_1}\right)^{\frac{1}{2}}.$$

Approximate Solutions. Except for the nonhomogeneous condition

$$(4) \qquad\qquad y(x,0) = f(x),$$

our boundary value problem is satisfied by any partial sum of the infinite series (4), Sec. 56,

$$(5) \qquad y_N(x,t) = \sum_{n=1}^{N} b_n \sin \frac{n\pi x}{c} \cos \frac{n\pi at}{c}.$$

In place of condition (4) this function satisfies the condition

$$(6) \qquad\qquad y_N(x,0) = \sum_{n=1}^{N} b_n \sin \frac{n\pi x}{c}.$$

The sum here is that of the first N terms of the Fourier sine series for $f(x)$, on the interval $(0,c)$. Since the odd periodic extension of f is continuous and f' is sectionally continuous, that series converges uniformly to $f(x)$, as shown in Chap. 5. Hence by taking N sufficiently large the sum $y_N(x,0)$ can be made to approximate $f(x)$ arbitrarily closely for all values of x on the interval $0 \leqq x \leqq c$.

The function $y_N(x,t)$, which is everywhere continuous together with all its partial derivatives, is therefore established as a solution of the approximating problem obtained by replacing condition (4) in the original problem by condition (6).

Corresponding approximations can be made to other problems. But a remarkable feature in the present case is that $y_N(x,t)$ never deviates from the actual displacement $y(x,t)$ by more than the greatest deviation of $y_N(x,0)$ from $f(x)$. This is true because

$$y_N = \frac{1}{2}\left[\sum_{n=1}^{N} b_n \sin \frac{n\pi(x-at)}{c} + \sum_{n=1}^{N} b_n \sin \frac{n\pi(x+at)}{c} \right]$$

and two sums here are those of the first N terms of the sine series for the odd periodic extension F of f, with arguments $x - at$ and $x + at$. But the greatest deviation of the first sum from $F(x - at)$, or of the second from $F(x + at)$, is the same as the greatest deviation of $y_N(x,0)$ from $f(x)$.

PROBLEMS

1. A string is stretched between the fixed points $(0,0)$ and $(1,0)$ and released at rest from the position $y = A \sin \pi x$. Find the formula for its subsequent displacements $y(x,t)$, and verify your result fully. Describe the motion of the string.   *Ans.* $y = A \sin \pi x \cos \pi at$.

2. Solve Problem 1 when the initial displacement there is changed to $y(x,0) = B \sin 2\pi x$.   *Ans.* $y = B \sin 2\pi x \cos 2\pi at$.

3. Show why the sum of the two functions $y(x,t)$ found in Problems 1 and 2 represents the displacements after the string is released at rest from the position $y = A \sin \pi x + B \sin 2\pi x$.

4. For the initially displaced string of length c considered in Secs. 56 and 57, note why the *frequency* ν of vibration in cycles per unit time has the value

$$\nu = \frac{a}{2c} = \frac{1}{2c}\sqrt{\frac{H}{\delta}}.$$

If the tension is 200 lb, the wt/ft 0.01 lb ($g\delta = 0.01$, $g = 32$), and the length 2 ft, show that $\nu = 200$ cycles/sec.

5. In Sec. 56, the position of the string at each instant can be shown graphically by moving the graph of the periodic function $\frac{1}{2}F(x)$ to the right with velocity a and an identical curve to the left at the same rate and adding ordinates, on the interval $0 \leqq x \leqq c$, of the two curves so obtained at the instant t. Show how this follows from formula (9).

6. Plot some positions of the plucked string considered in Sec. 18 by the method described in Problem 5 to verify that the string assumes such positions as indicated by the dotted lines in Fig. 13.

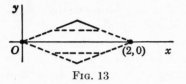

FIG. 13

7. Write the boundary value problem (1) to (3), Sec. 56, in terms of the two independent variables x and τ, where $\tau = at$, to show that the problem in y as a function of x and τ does not involve the constant a. Thus, without solving the problem, deduce that the solution has the form $y = \phi(x,\tau) = \phi(x,at)$ and hence that the relation (2), Sec. 57, follows.

58. Prescribed Initial Velocity. If, initially, the string has some prescribed distribution of velocities $g(x)$ parallel to the y axis in its position of equilibrium $y = 0$, the boundary value problem for the displacements $y(x,t)$ becomes

$$(1) \qquad y_{tt}(x,t) = a^2 y_{xx}(x,t) \qquad\qquad (0 < x < c,\, t > 0),$$
$$(2) \qquad y(0,t) = 0, \qquad y(c,t) = 0 \qquad\qquad (t \geqq 0),$$
$$(3) \qquad y(x,0) = 0, \qquad y_t(x,0) = g(x) \qquad (0 < x < c).$$

In case the xy plane, with the string lying on the x axis, is moving parallel to the y axis and is brought to rest at the instant $t = 0$, then g is a constant. The hammer action in a piano may produce approximately a uniform initial velocity over a short span of a piano wire, in which case g may be considered to be a step function.

As before, we find the functions of type $X(x)T(t)$ that satisfy all *homogeneous* conditions in the problem (1) to (3). The Sturm-Liouville problem in X is the same as the one found when the initial displacement was prescribed; thus $\lambda = n^2\pi^2/c^2$ and $X = \sin(n\pi x/c)$. The conditions on T become

$$T''(t) + \lambda a^2 T(t) = 0, \qquad T(0) = 0 \quad (\lambda = n^2\pi^2/c^2);$$

hence $T = \sin(n\pi at/c)$. To satisfy the nonhomogeneous condition as well, we write, formally,

$$y(x,t) = \sum_{n=1}^{\infty} A_n \sin \frac{n\pi x}{c} \sin \frac{n\pi a t}{c},$$

$$(4) \qquad y_t(x,0) = \sum_{n=1}^{\infty} \frac{n\pi a}{c} A_n \sin \frac{n\pi x}{c} = g(x) \quad (0 < x < c).$$

Assuming that g and g' are sectionally continuous, the series in formula (4) is the Fourier sine series that represents $g(x)$ on the interval $(0,c)$ if $n\pi a A_n/c = b_n$, where

$$(5) \qquad b_n = \frac{2}{c} \int_0^c g(x) \sin \frac{n\pi x}{c}\, dx.$$

Thus $A_n = b_n c/(n\pi a)$ and

$$(6) \qquad y(x,t) = \frac{c}{\pi a} \sum_{n=1}^{\infty} \frac{b_n}{n} \sin \frac{n\pi x}{c} \sin \frac{n\pi a t}{c}.$$

We can sum the series here by first writing

$$y_t(x,t) = \sum_{n=1}^{\infty} b_n \sin \frac{n\pi x}{c} \cos \frac{n\pi a t}{c}$$

$$= \frac{1}{2} G(x + at) + \frac{1}{2} G(x - at),$$

where G is the odd periodic extension, with period $2c$, of the given function g (compare Sec. 56). Then, since $y(x,0) = 0$,

$$y(x,t) = \frac{1}{2} \int_0^t G(x + a\tau)\, d\tau + \frac{1}{2} \int_0^t G(x - a\tau)\, d\tau$$

$$= \frac{1}{2a} \int_x^{x+at} G(\xi)\, d\xi - \frac{1}{2a} \int_x^{x-at} G(\xi)\, d\xi.$$

In terms of the function

$$(7) \qquad H(x) = \int_0^x G(\xi)\, d\xi \qquad (-\infty < x < \infty),$$

$$(8) \qquad y(x,t) = \frac{1}{2a} [H(x + at) - H(x - at)].$$

The verification of the solution in the form (8), and the identification of the functions defined by (6) and (8), are left to the problems.

Superposition of Solutions. If the string is given both an initial displacement and initial velocity,

$$(9) \qquad y(x,0) = f(x), \qquad y_t(x,0) = g(x),$$

the displacements $y(x,t)$ can be written as the superposition of the solution (8) above and the solution (9), Sec. 56, namely,

$$(10) \quad y(x,t) = \frac{1}{2}\left[F(x + at) + F(x - at)\right] + \frac{1}{2a}\left[H(x + at)\right.$$
$$\left. - H(x - at)\right].$$

Note that both terms satisfy the homogeneous conditions (1) and (2), while their sum clearly satisfies the nonhomogeneous conditions (9).

In general the solution of a linear problem containing more than one nonhomogeneous condition can be written as a sum of solutions of problems each of which contains only one nonhomogeneous condition. The resolution of the original problem in this way, though not an essential step, often simplifies the process of solving the problem.

59. Nonhomogeneous Differential Equations. The substitution of a new unknown function sometimes reduces a nonhomogeneous partial differential equation to one that is homogeneous. Then if variables can be separated and if the new two-point boundary conditions can be made homogeneous by properly selecting the new function so that a Sturm-Liouville problem arises, our method of superposition may be effective.

To illustrate, consider the displacements in a stretched string upon which an external force per unit length acts, parallel to the y axis. Let that force be proportional to the distance from one end, and let the initial displacement and velocity be zero. Units for x and t can be chosen so that the problem becomes

$$(1) \qquad\qquad y_{tt}(x,t) = y_{xx}(x,t) + Ax \quad (0 < x < 1, t > 0),$$
$$(2) \qquad y(0,t) = y(1,t) = y(x,0) = y_t(x,0) = 0.$$

In terms of the new function Y, where

$$y(x,t) = Y(x,t) + \psi(x)$$

and ψ is yet to be determined, equation (1) becomes

$$Y_{tt}(x,t) = Y_{xx}(x,t) + \psi''(x) + Ax.$$

This will be homogeneous if

(3) $\psi''(x) = -Ax.$

The two-point boundary conditions on Y become

$$Y(0,t) + \psi(0) = 0, \qquad Y(1,t) + \psi(1) = 0;$$

they are homogeneous if

(4) $\psi(0) = 0, \qquad \psi(1) = 0.$

From conditions (3) and (4) we find that

(5) $\psi(x) = \dfrac{A}{6} x(1 - x^2)$ $(0 \leqq x \leqq 1).$

The boundary value problem in Y now consists of the conditions $Y_{tt} = Y_{xx}$, $Y(0,t) = Y(1,t) = 0$, and

$$Y(x,0) = -\psi(x), \qquad Y_t(x,0) = 0.$$

This is a special case of the problem solved in Sec. 56. The solution of our problem therefore can be written

(6) $y(x,t) = \psi(x) - \tfrac{1}{2}[\Psi(x - t) + \Psi(x + t)],$

where $\Psi(x)$ is defined for all finite x as the odd periodic extension of ψ, with period 2.

60. Elastic Bar. A cylindrical bar of natural length c is initially stretched by an amount bc (Fig. 14) and at rest. The

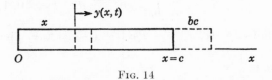

Fig. 14

initial longitudinal displacements of its sections are then proportional to the distance from the fixed end. At the instant $t = 0$ both ends are released and *kept free*. The boundary value problem in the *longitudinal displacements* $y(x,t)$ is (Sec. 5)

(1) $y_{tt}(x,t) = a^2 y_{xx}(x,t)$ $(0 < x < c, t > 0; a^2 = E/\delta),$

(2) $y_x(0,t) = 0, \qquad y_x(c,t) = 0$ $(t > 0),$

(3) $y(x,0) = bx, \qquad y_t(x,0) = 0$ $(0 < x < c).$

The homogeneous two-point boundary conditions (2) state that the force per unit area $E \, \partial y/\partial x$ on the end sections is zero.

Functions $X(x)T(t)$ satisfy all the homogeneous conditions above when X is an eigenfunction of the problem

(4) $\qquad X''(x) + \lambda X(x) = 0, \qquad X'(0) = X'(c) = 0,$

and when, for the same eigenvalue λ,

(5) $\qquad T''(t) + \lambda a^2 T(t) = 0, \qquad T'(0) = 0.$

The Sturm-Liouville problem (4) generates the eigenfunctions used in the Fourier cosine series on the interval $(0,c)$. The eigenvalues, all real, are $\lambda_0 = 0$, $\lambda_n = n^2\pi^2/c^2$, $n = 1, 2, \ldots$; also $X_0(x) = 1$, $T_0(t) = 1$, and

$$X_n(x) = \cos (n\pi x/c), \qquad T_n(t) = \cos (n\pi a t/c).$$

Formally then, the generalized linear combination

$$y(x,t) = \frac{1}{2} a_0 + \sum_{n=1}^{\infty} a_n \cos \frac{n\pi x}{c} \cos \frac{n\pi a t}{c}$$

satisfies all the conditions (1) to (3) provided that

(6) $\qquad\qquad bx = \dfrac{1}{2} a_0 + \displaystyle\sum_{n=1}^{\infty} a_n \cos \dfrac{n\pi x}{c} \qquad (0 < x < c).$

The function bx is such that it is represented by this Fourier cosine series on the interval $0 \leq x \leq c$, where

$$a_n = \frac{2b}{c} \int_0^c x \cos \frac{n\pi x}{c} \, dx \quad (n = 0, 1, 2, \ldots).$$

We find that $a_0 = bc$, $a_n = -2bc[1 - (-1)^n]/(n\pi)^2$; hence

(7) $y(x,t) = \dfrac{1}{2} bc$

$$- \frac{4bc}{\pi^2} \sum_{n=1}^{\infty} \frac{1}{(2n-1)^2} \cos \frac{(2n-1)\pi x}{c} \cos \frac{(2n-1)\pi a t}{c}.$$

If $P(x)$ denotes the *even* periodic extension, with period $2c$, of the function bx $(0 \leq x \leq c)$, formula (7) can be written

(8) $\qquad\qquad y(x,t) = \tfrac{1}{2}[P(x + at) + P(x - at)].$

This reduction, and the verification of the solution in form (8), are left as exercises.

PROBLEMS

1. Show that, for each fixed x, the displacements y given by formula (10), Sec. 58, are periodic functions of t with period $2c/a$.

2. Show that the motion of each cross section of the elastic bar treated in Sec. 60 is periodic in t with period $2c/a$.

3. A string stretched between the points $(0,0)$ and $(\pi,0)$ is initially straight with velocity $y_t(x,0) = b \sin x$. Write the boundary value problem in $y(x,t)$, solve it, and verify the solution.

$$Ans. \quad y = a^{-1}b \sin x \sin at.$$

4. A wire stretched between the points $(0,0)$ and $(1,0)$ of a horizontal x axis is initially at rest along that axis, its subsequent motion being caused by the force of gravity. If the y axis is positive upwards, the equation of motion is $y_{tt} = a^2 y_{xx} - g$, where g is the acceleration of gravity. Write the boundary value problem in the vertical displacements $y(x,t)$ and solve it. Show that

$$y(x,t) = \psi(x) - \tfrac{1}{2}[\Psi(x - at) + \Psi(x + at)],$$

where the function Ψ is the odd periodic extension, with period 2, of the function

$$\psi(x) = \frac{g}{2a^2} x(x - 1) \qquad (0 \leqq x \leqq 1).$$

5. In Sec. 58, when all points of the string have the same initial velocity, $g(x) = v_0$, display graphically the periodic functions $G(x)$ and $H(x)$ and note that the instantaneous positions of the string given by formula (8) are broken lines similar to the positions shown in Fig. 13.

6. Verify the solution (8), Sec. 58. Note that the function H defined by equation (7) is continuous, even, and periodic and that $H'(x) = G(x)$ wherever G is continuous.

7. In Sec. 58 we noted that

$$G(r) = \sum_{n=1}^{\infty} b_n \sin \frac{n\pi r}{c} \qquad (-\infty < r < \infty).$$

Integrate that series (Sec. 48) to show that

$$H(x) = \frac{c}{\pi} \sum_{n=1}^{\infty} \frac{1}{n} b_n \left(1 - \cos \frac{n\pi x}{c}\right) \qquad (-\infty < x < \infty)$$

and hence that the function (8) is represented by the series (6).

8. In Sec. 60, reduce the representation (7) of the function y to the form (8), and verify the solution.

9. From the formula (8), Sec. 60, show that $y(0,t) = P(at)$; hence that the end $x = 0$ of the bar moves with constant velocity ab during the half period $0 < t < c/a$ and with velocity $-ab$ during the next half period.

10. If the function y that satisfies conditions (1) to (3), Sec. 60, is interpreted to represent transverse displacements in a stretched string, where $a^2 = H/\delta$, describe the initial and end conditions for the string. [Note that conditions (2) imply that no vertical force acts at the ends, as if the ends were looped about perfectly smooth rods along the lines $x = 0$ and $x = c$.]

61. Temperatures in a Bar. The lateral surface of a solid right cylinder of length π is insulated. The initial temperature distribution within the bar is a prescribed function f of the distance x from the base $x = 0$. At the instant $t = 0$ the temperature of both bases $x = 0$ and $x = \pi$ is brought to zero and kept at that value (Fig. 15). If no heat is generated in the solid,

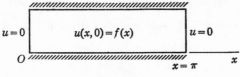

FIG. 15

the temperatures should be the values of a function $u(x,t)$ that satisfies the heat equation

$$(1) \qquad u_t(x,t) = ku_{xx}(x,t) \qquad (0 < x < \pi, t > 0)$$

and the boundary conditions

$$(2) \qquad u(+0,t) = 0, \qquad u(\pi - 0, t) = 0 \qquad (t > 0),$$
$$(3) \qquad u(x,+0) = f(x) \qquad (0 < x < \pi).$$

The constant k is the diffusivity of the material. We have written the boundary conditions as limits in order to indicate continuity properties that should be satisfied by u. The problem is also that of determining temperatures $u(x,t)$ in a slab bounded by the planes $x = 0$ and $x = \pi$, initially at temperature $f(x)$, with its faces kept at temperature zero.

By separation of variables we find that $X(x)T(t)$ satisfies the homogeneous conditions (1) and (2) if

(4) $X''(x) + \lambda X(x) = 0,$ $X(0) = X(\pi) = 0,$
(5) $T'(t) = -\lambda k T(t).$

The Sturm-Liouville problem (4) has the eigenvalues $\lambda = n^2$ and eigenfunctions $\sin nx$ $(n = 1, 2, \ldots)$. Corresponding functions T, from equation (5), are $\exp(-n^2kt)$. Formally then, the function

(6) $u(x,t) = \sum_{n=1}^{\infty} b_n \exp(-n^2kt) \sin nx$

satisfies all conditions, including (3), if

(7) $f(x) = \sum_{n=1}^{\infty} b_n \sin nx$ $(0 < x < \pi).$

We assume that f and f' are sectionally continuous. Then f is represented by its Fourier sine series (7), where

(8) $b_n = \frac{2}{\pi} \int_0^\pi f(x) \sin nx \, dx$ $(n = 1, 2, \ldots).$

Formula (6), with coefficients b_n defined by equation (8), is our formal solution of the boundary value problem.

a. *Verification.* We have seen earlier that $b_n \to 0$ as $n \to \infty$; hence those coefficients are bounded for all n, $|b_n| < M$, where M is some constant. Whenever $t \geqq t_0$, where t_0 is a positive constant,

$$|b_n \exp(-n^2kt) \sin nx| < M \exp(-n^2kt_0).$$

Since the infinite series with constant terms $\exp(-n^2kt_0)$ converges, according to the ratio test, the Weierstrass test ensures the uniform convergence of series (6) with respect to x and t when $0 \leqq x \leqq \pi$, $t \geqq t_0 > 0$. The terms of that series are continuous functions; hence series (6) converges to a continuous function $u(x,t)$ when $t \geqq t_0$, that is, whenever $t > 0$ since t_0 is an arbitrary positive number. In particular, $u(+0,t) = u(0,t)$ when $t > 0$, and since $u(0,t) = 0$, the first of conditions (2) is satisfied by the function u defined by formula (6). Similarly, the second of those conditions is satisfied.

The series with terms $n \exp(-n^2kt_0)$, or $n^2 \exp(-n^2kt_0)$, also converges. Hence series (6) can be differentiated twice with respect to x and once with respect to t, when $t > 0$, because the series of derivatives converge uniformly when $t \geq t_0$. But the terms of series (6) satisfy the heat equation (1), and so the sum $u(x,t)$ of the series satisfies that homogeneous differential equation.

It remains to show that u satisfies the initial condition (3). This can be done with the aid of a convergence test due to Abel,[1] derived in Chap. 10. For each fixed x $(0 < x < \pi)$ the series with terms $b_n \sin nx$ converges to $f(x)$; at a point of discontinuity we define $f(x)$ as the mean of the values $f(x + 0)$ and $f(x - 0)$. According to Abel's test, the new series formed by multiplying the terms of a convergent series by corresponding members of a bounded sequence of functions of t, such as $\exp(-n^2kt)$, whose values never increase with n, is uniformly convergent with respect to t. Our series (6) therefore converges uniformly with respect to t when $t \geq 0$.

It follows that our function u is continuous in t $(t \geq 0)$; thus $u(x,+0) = u(x,0)$. Since $u(x,0) = f(x)$, condition (3) is satisfied. The function u defined by (6) is now established as a solution.

b. *Uniqueness.* Consider the somewhat special case in which f is continuous $(0 \leq x \leq \pi)$, $f(0) = f(\pi) = 0$, and f' is sectionally continuous so that the Fourier sine series (7) converges uniformly to $f(x)$. Then Abel's test shows uniform convergence of series (6) with respect to x and t together in the region $0 \leq x \leq \pi, t \geq 0$. Consequently our function u is continuous in that region. As before, u_t, u_x, and u_{xx} are continuous when $t > 0$ and $0 \leq x \leq \pi$.

A simple uniqueness theorem for solutions of such boundary value problems is proved in Chap. 10. It shows that not more than one function $U(x,y,z,t)$ can satisfy the heat equation throughout the interior of a bounded closed region R in space, such as our cylindrical region, and meet the following requirements: U is to be a continuous function of x, y, z, and t together when $t \geq 0$ and when (x,y,z) is in R, where R includes its boundary surface; the derivatives of U that appear in the heat equation are continuous in R when $t > 0$; U is prescribed on part of the boundary and the normal derivative of U on the rest when $t > 0$.

Our function $u(x,t)$, which is independent of y and z, satisfies all those requirements, including conditions (1) to (3). For each fixed t the gradient of u is parallel to the x axis, so the normal derivative of u at the lateral surface of the cylinder, or the flux

[1] Niels H. Abel (pronounced Ah-bel), Norwegian, 1802–1829.

of heat across that surface, is zero.   Hence formula (6) represents
the solution of our problem that satisfies the continuity conditions
stated.   We have now justified the assumption that u is a function of x and t only.

62. Other Boundary Conditions.   We point out modifications
of the procedure that are useful when the slab $0 \leq x \leq \pi$, or
the bar with insulated lateral surface, is subjected to other
simple thermal conditions at its boundary surfaces $x = 0$ and
$x = \pi$.   In each case the temperature function is to satisfy the
heat equation in x and t within the slab,

$$(1) \qquad u_t(x,t) = ku_{xx}(x,t) \qquad (0 < x < \pi, t > 0).$$

a. One Face at Temperature u_0.   If the slab is initially at temperature zero throughout and the face $x = 0$ is kept at that
temperature while the face $x = \pi$ is kept at a constant temperature u_0 when $t > 0$, then

$$(2) \qquad u(0,t) = 0, \qquad u(\pi,t) = u_0, \qquad u(x,0) = 0.$$

The boundary value problem consisting of equations (1) and
(2) is not in proper form for the method of separating variables
because one of the two-point boundary conditions is not homogeneous.   If we write

$$v(x,t) = u(x,t) - \psi(x),$$

we find that when $\psi(x) = u_0 x/\pi$ (the steady-state temperature
in this slab), the problem in v becomes

$$(3) \quad v_t = kv_{xx}, \qquad v(0,t) = v(\pi,t) = 0, \qquad v(x,0) = -\psi(x).$$

This is a special case of the problem solved in Sec. 61, where
$f(x) = -u_0 x/\pi$.   Thus we obtain the formula

$$(4) \qquad u(x,t) = \frac{u_0}{\pi}\left[x + 2\sum_{n=1}^{\infty}\frac{(-1)^n}{n}\exp\left(-n^2kt\right)\sin nx\right].$$

b. Insulated Faces.   The flux of heat across the faces $x = 0$
and $x = \pi$ is the value of $-Ku_x$ at those faces, where K is the
thermal conductivity.   If both faces of the slab are insulated and
the initial temperature is $f(x)$, the boundary value problem consists of equation (1) and the conditions

$$(5) \qquad u_x(0,t) = u_x(\pi,t) = 0, \qquad u(x,0) = f(x) \quad (0 < x < \pi).$$

Separation of variables now leads to the Sturm-Liouville problem associated with the Fourier cosine series,

$$X''(x) + \lambda X(x) = 0, \qquad X'(0) = X'(\pi) = 0,$$

and to the equation $T'(t) = -\lambda k T(t)$. The temperature function is found to be

$$(6) \qquad u(x,t) = \tfrac{1}{2}a_0 + \sum_{n=1}^{\infty} a_n \exp(-n^2 kt) \cos nx,$$

where

$$a_n = \frac{2}{\pi} \int_0^{\pi} f(x) \cos nx \, dx \quad (n = 0, 1, 2, \ldots).$$

c. One Face Insulated. If the face $x = \pi$ is insulated while the

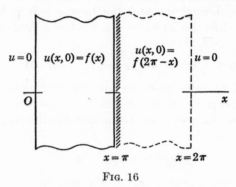

FIG. 16

face $x = 0$ is kept at temperature zero (Fig. 16), then

$$(7) \qquad u(0,t) = u_x(\pi,t) = 0, \qquad u(x,0) = f(x) \quad (0 < x < \pi),$$

where $f(x)$ is the initial temperature. Upon separating variables we obtain the Sturm-Liouville problem

$$X''(x) + \lambda X(x) = 0, \qquad X(0) = 0, \qquad X'(\pi) = 0$$

and the equation $T'(t) = -\lambda k T(t)$. The eigenvalues and normalized eigenfunctions of that problem are found to be

$$(8) \quad \lambda_n = \frac{1}{4}(2n-1)^2, \qquad \phi_n(x) = \sqrt{\frac{2}{\pi}} \sin \frac{(2n-1)x}{2}$$

$$(n = 1, 2, \ldots).$$

Note that the orthogonal sine functions here are *not* those used in the Fourier sine series for the interval $(0,\pi)$.

Our generalized linear combination of functions of type XT, written in terms of the functions ϕ_n, is

$$(9) \quad u(x,t) = \sum_{n=1}^{\infty} c_n \exp\left[-\frac{1}{4}(2n - 1)^2 kt \right] \sqrt{\frac{2}{\pi}} \sin \frac{(2n - 1)x}{2}.$$

It satisfies the condition $u(x,0) = f(x)$ if

$$(10) \qquad f(x) = \sum_{n=1}^{\infty} c_n \sqrt{\frac{2}{\pi}} \sin \frac{(2n - 1)x}{2} \qquad (0 < x < \pi).$$

Series (10) is the generalized Fourier series corresponding to f with respect to the eigenfunctions ϕ_n if

$$(11) \quad c_n = \int_0^{\pi} f(x)\phi_n(x)\, dx = \sqrt{\frac{2}{\pi}} \int_0^{\pi} f(x) \sin \frac{(2n - 1)x}{2}\, dx.$$

If we assume that the representation (10) is valid, as proved in references to Sturm-Liouville theory cited in Sec. 32, formula (9) with coefficients c_n defined by (11) represents a formal solution of our problem.

But we can actually establish the special representation (10) with the aid of Fourier sine series on the larger interval $(0,2\pi)$. Let the given function f and its derivative f' be sectionally continuous on the interval $(0,\pi)$, then let f be defined on $(\pi,2\pi)$ so that its graph is symmetric to the line $x = \pi$; that is,

$$(12) \qquad f(x) = f(2\pi - x) \qquad \text{when } \pi < x < 2\pi.$$

This procedure is suggested by considering the temperature problem in the slab extended to the plane $x = 2\pi$ with that plane kept at temperature zero (Fig. 16). When the initial temperature distribution has the symmetry property (12), no heat should flow across the midsection $x = \pi$, so the temperature function u should be the one sought when $0 \leq x \leq \pi$. The new problem is essentially the one solved in Sec. 61 except that now the sine series on the interval $(0,2\pi)$, for the symmetric extension of f, is to be used.

That sine series representation on $(0,c)$, where $c = 2\pi$, is

$$(13) \qquad f(x) = \sum_{n=1}^{\infty} b_n \sin \frac{nx}{2} \qquad (0 < x < 2\pi),$$

where $b_n = \dfrac{1}{\pi} \displaystyle\int_0^{2\pi} f(x) \sin \dfrac{nx}{2}\, dx$

$\qquad\quad = \dfrac{1}{\pi} \displaystyle\int_0^{\pi} f(\xi) \sin \dfrac{n\xi}{2}\, d\xi + \dfrac{1}{\pi} \displaystyle\int_{\pi}^{2\pi} f(2\pi - x) \sin \dfrac{nx}{2}\, dx.$

In the last integral we substitute ξ for $2\pi - x$ and note that $\sin (nx/2)$ then becomes $-(-1)^n \sin (n\xi/2)$.   Thus

$$b_n = \frac{1}{\pi} [1 - (-1)^n] \int_0^{\pi} f(\xi) \sin \frac{n\xi}{2}\, d\xi;$$

that is,

(14) $\qquad b_{2n-1} = \dfrac{2}{\pi} \displaystyle\int_0^{\pi} f(\xi) \sin \dfrac{(2n - 1)\xi}{2}\, d\xi, \qquad b_{2n} = 0.$

The representation (13) with coefficients (14) is therefore valid for the given function f on the interval $(0,\pi)$.   It is the same as the representation (10) with coefficients (11).

PROBLEMS[1]

1. The faces $x = 0$ and $x = c$ of a slab which is initially at temperatures $f(x)$ are kept at temperature zero (compare Sec. 61).   Derive the temperature formula

$$u(x,t) = \frac{2}{c} \sum_{n=1}^{\infty} \exp\left(-\frac{n^2\pi^2kt}{c^2}\right) \sin \frac{n\pi x}{c} \int_0^c f(\xi) \sin \frac{n\pi \xi}{c}\, d\xi.$$

2. When the initial temperature distribution is uniform over the slab in Problem 1, $f(x) = u_0$, write the formula for $u(x,t)$ and for the flux $-Ku_x(x_0,t)$ across a plane $x = x_0$ $(0 \leqq x_0 \leqq c,\ t > 0)$ and note that no heat flows across the center plane $x = c/2$.

3. If $f(x) = \sin (\pi x/c)$ in Problem 1, find $u(x,t)$ and verify your result fully. $\qquad\qquad\qquad\qquad$ *Ans.* $u = \exp (-\pi^2kt/c^2) \sin (\pi x/c)$.

4. (a) In Problem 1, if $f(x) = A$ when $0 < x < \frac{1}{2}c$ and $f(x) = 0$ when $\frac{1}{2}c < x < c$, show that

$$u(x,t) = \frac{4A}{\pi} \sum_{n=1}^{\infty} \frac{\sin^2 (n\pi/4)}{n} \exp\left(-\frac{n^2\pi^2kt}{c^2}\right) \sin \frac{n\pi x}{c}.$$

(b) Two slabs of iron ($k = 0.15$ centimeter-gram-second unit), each

---

[1] Only formal solutions of the boundary value problems here and in the sets of problems to follow are expected, unless the problem states that the solution found is to be fully verified.   Partial verification is often easy and helpful.

20 cm thick, one at temperature 100°C and the other at 0°C throughout, are placed face to face in perfect contact, and their outer faces are kept at 0°C. Compute the temperature at their common face 10 min after contact was made to show that it is approximately 36°C.

(c) If the slabs in part (b) are made of concrete ($k = 0.005$ cgs unit), show that it takes 5 hr for the common face to reach that temperature 37°C. [Note that $u(x,t)$ depends on the product kt.]

5. The faces $x = 0$ and $x = c$ of a slab which is initially at temperatures $f(x)$ are insulated (compare Sec. 62b). Derive the temperature formula

$$u(x,t) = \frac{1}{2} a_0 + \sum_{n=1}^{\infty} a_n \exp\left(-\frac{n^2\pi^2 kt}{c^2}\right) \cos\frac{n\pi x}{c},$$

where
$$a_n = \frac{2}{c} \int_0^c f(x) \cos\frac{n\pi x}{c}\, dx \quad (n = 0, 1, 2, \ldots).$$

Note that if $f(x) = u_0$, a constant, then $u(x,t) = u_0$.

6. Establish the result in Problem 5 (compare Sec. 61a).

7. Let $v(x,t)$ denote temperatures in a slender wire with its axis along the x axis. Variations of the temperature over each cross section are

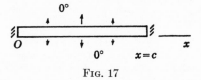

Fig. 17

neglected. At the lateral surface the *linear law of surface heat transfer* between the wire and its surroundings is assumed to apply (Problem 7, Sec. 7). Let the surroundings be at temperature zero; then

$$v_t(x,t) = kv_{xx}(x,t) - hv(x,t) \quad (h \text{ constant}, h > 0).$$

If the ends $x = 0$ and $x = c$ are insulated (Fig. 17) and the initial temperature distribution is $f(x)$, solve the boundary value problem for v and show that
$$v(x,t) = e^{-ht}u(x,t),$$

where u is the temperature function found in Problem 5.

8. Use the substitution $u(x,t) = e^{ht}v(x,t)$ to reduce the boundary value problem in Problem 7 to the one in Problem 5.

9. If the ends of the wire in Problem 7 are not insulated, but kept at temperature zero instead, find the temperature function.

10. Heat is generated at a constant rate uniformly throughout a slab, initially at temperatures $f(x)$, whose faces $x = 0$ and $x = \pi$ are kept at temperature zero. Then the heat equation has the form (Sec. 7)

$$u_t(x,t) = ku_{xx}(x,t) + C,$$

where C is a positive constant. Solve the boundary value problem to obtain the temperature formula

$$u(x,t) = \frac{Cx}{2k}(\pi - x) + \sum_{n=1}^{\infty} b_n \exp\,(-n^2kt)\,\sin\,nx,$$

where $\qquad b_n = \dfrac{2}{\pi}\displaystyle\int_0^{\pi}\left[\dfrac{Cx}{2k}(x - \pi) + f(x)\right]\sin\,nx\,dx.$

Also, note the result when $f(x) = \frac{1}{2}Cx(\pi - x)/k$.

11. Solve this boundary value problem for $u(x,t)$:

$$u_t(x,t) = u_{xx}(x,t) - hu(x,t) \quad (0 < x < \pi, \, t > 0; \, h \text{ constant}, \, h > 0),$$
$$u(0,t) = 0, \qquad u(\pi,t) = 1 \qquad\qquad\qquad (t > 0),$$
$$u(x,0) = 0 \qquad\qquad\qquad\qquad (0 < x < \pi).$$

Ans. $u(x,t) = \dfrac{\sinh x \sqrt{h}}{\sinh \pi \sqrt{h}} + \dfrac{2}{\pi}\,e^{-ht}\displaystyle\sum_{n=1}^{\infty}\dfrac{n(-1)^n}{n^2 + h}\exp\,(-n^2t)\,\sin\,nx.$

12. Use superposition of known solutions (Secs. 61 and 62) to write a formula for temperatures in a slab whose faces $x = 0$ and $x = \pi$ are kept at temperatures 0 and u_0, respectively, and whose initial temperature distribution is $f(x)$.

13. The face $x = 0$ of a slab is kept at temperature zero while heat is supplied or extracted at the other face $x = \pi$ at a constant rate per unit area so that $u_x(\pi,t) = A$. If the slab is initially at temperature zero and if the unit of time is chosen so that we can write $k = 1$, derive the temperature formula

$$u(x,t) = Ax + \frac{8A}{\pi}\sum_{n=1}^{\infty}\frac{(-1)^n}{(2n - 1)^2}\exp\left[-\frac{(2n - 1)^2t}{4}\right]\sin\frac{(2n - 1)x}{2}$$

with the aid of the series representation (10), Sec. 62.

14. Let f and f' be sectionally continuous on an interval $(0,c)$. By the process used in Sec. 62c, establish the representation

$$f(x) = \frac{2}{c}\sum_{n=1}^{\infty}\sin\frac{(2n - 1)\pi x}{2c}\int_0^c f(\xi)\sin\frac{(2n - 1)\pi \xi}{2c}\,d\xi \quad (0 < x \leqq c)$$

of f by the series of the eigenfunctions of the Sturm-Liouville problem

$$X''\,(x) + \lambda X(x) = 0, \qquad X(0) = 0, \qquad X'\,(c) = 0.$$

63. A Dirichlet Problem. Let u be a harmonic function in the interior of a rectangle in the xy plane

$$(1) \qquad u_{xx}(x,y) + u_{yy}(x,y) = 0 \quad (0 < x < a, 0 < y < b)$$

with these values prescribed on the boundary (Fig. 18):

$$(2) \qquad u(0,y) = 0, \qquad u(a,y) = 0 \qquad (0 < y < b),$$
$$(3) \qquad u(x,0) = f(x), \qquad u(x,b) = 0 \qquad (0 < x < a).$$

Separation of variables leads to the Sturm-Liouville problem

$$(4) \qquad X''(x) + \lambda X(x) = 0, \qquad X(0) = X(a) = 0,$$

where $\lambda = n^2\pi^2 a^{-2}$ and $X_n = \sin(n\pi x/a)$, $n = 1, 2, \ldots$, and

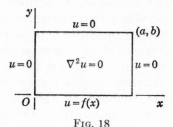

Fig. 18

to the conditions

$$(5) \qquad Y''(y) - n^2\pi^2 a^{-2} Y(y) = 0, \qquad Y(b) = 0.$$

Thus Y can be written as $\sinh[n\pi a^{-1}(b - y)]$. The function

$$(6) \qquad u(x,y) = \sum_{n=1}^{\infty} A_n \sinh \frac{n\pi(b - y)}{a} \sin \frac{n\pi x}{a}$$

formally satisfies all conditions (1) to (3) if

$$(7) \qquad f(x) = \sum_{n=1}^{\infty} A_n \sinh \frac{n\pi b}{a} \sin \frac{n\pi x}{a} \qquad (0 < x < a).$$

We assume that f and f' are sectionally continuous. Then formula (7) is the Fourier sine series representation of f on the interval $(0,a)$ provided $A_n \sinh(n\pi b/a) = b_n$, where

$$b_n = \frac{2}{a} \int_0^a f(x) \sin \frac{n\pi x}{a} \, dx.$$

The function defined by equation (6) with coefficients

$$(8) \qquad A_n = \frac{2}{a \sinh{(n\pi b/a)}} \int_0^a f(x) \sin\frac{n\pi x}{a}\, dx$$

is therefore our formal solution. The result can be established by the method used in Sec. 61. We defer the verification of the solution to Chap. 10, where uniqueness is also considered.

If y is replaced by the new variable $b - y$ in the above problem and solution, and f by g, the nonhomogeneous condition satisfied by u becomes $u(x,b) = g(x)$. The interchange of x and y places the nonhomogeneous conditions at the boundaries $x = 0$ or $x = a$. Superposition of the four solutions then gives the harmonic function whose values are prescribed as a function of position along the entire boundary of the rectangle.

From the conditions (1) to (3) we note that $u(x,y)$ represents the *steady-state temperatures* in a rectangular plate with insulated faces when $u = f(x)$ on the edge $y = 0$, and $u = 0$ on the other three edges. The function u also represents the *electrostatic potential* in the space bounded by the planes $x = 0$, $x = a$, $y = 0$, and $y = b$ when the space is free of charges and the plane boundaries are kept at potentials given by conditions (2) and (3). If f is continuous and $f(0) = f(a) = 0$, $u(x,y)$ represents the *static transverse displacements in a membrane* stretched over a rectangular frame after the supporting side $y = 0$ is displaced, $u(x,0) = f(x)$.

64. Fourier Series in Two Variables. Let $z(x,y,t)$ denote the transverse displacement at each point (x,y) at time t in a membrane stretched across a rigid square frame in the xy plane. To simplify the notation we select the origin and the point (π,π) as ends of a diagonal of the frame. If the membrane is released at rest with a given initial displacement $f(x,y)$, where f is continuous and vanishes on the boundary of the square, then

$$(1) \qquad z_{tt} = a^2(z_{xx} + z_{yy}) \quad (0 < x < \pi, 0 < y < \pi, t > 0),$$
$$(2) \qquad z(0,y,t) = z(\pi,y,t) = z(x,0,t) = z(x,\pi,t) = 0,$$
$$(3) \qquad z(x,y,0) = f(x,y), \qquad z_t(x,y,0) = 0$$
$$(0 \le x \le \pi, 0 \le y \le \pi).$$

Let us assume that $f_x(x,y)$ and $f_y(x,y)$ are also continuous.

Functions of type XYT satisfy equation (1) if

$$\frac{X''(x)}{X(x)} = \frac{T''(t)}{a^2 T(t)} - \frac{Y''(y)}{Y(y)}.$$

Thus $X''/X = -\lambda$ and $Y''/Y = T''/(a^2T) + \lambda$, where λ is a constant; also $Y''/Y = -\mu$ and $T''/(a^2T) + \lambda = -\mu$, where the constant μ is independent of λ. Separation of variables in all homogeneous conditions therefore leads to the two Sturm-Liouville problems

$$X''(x) + \lambda X(x) = 0, \qquad X(0) = X(\pi) = 0,$$
$$Y''(y) + \mu Y(y) = 0, \qquad Y(0) = Y(\pi) = 0$$

and to the conditions

$$T''(t) + a^2(\lambda + \mu)T(t) = 0, \qquad T'(0) = 0.$$

Consequently $\lambda = m^2$ $(m = 1, 2, \ldots)$, $X = \sin mx$, $\mu = n^2$ $(n = 1, 2, \ldots)$, $Y = \sin ny$, and $T = \cos (at \sqrt{m^2 + n^2})$.

The formal solution of our problem is then

$$(4) \quad z(x,y,t) = \sum_{m=1}^{\infty} \sum_{n=1}^{\infty} A_{mn} \sin mx \sin ny \cos (at \sqrt{m^2 + n^2})$$

if the coefficients A_{mn} can be determined so that

$$(5) \qquad f(x,y) = \sum_{m=1}^{\infty} \sum_{n=1}^{\infty} A_{mn} \sin mx \sin ny$$

when $0 \leqq x \leqq \pi$ and $0 \leqq y \leqq \pi$. By grouping terms in this double sine series so as to display the total coefficient of $\sin mx$ for each m we can write, formally,

$$(6) \qquad f(x,y) = \sum_{m=1}^{\infty} \left(\sum_{n=1}^{\infty} A_{mn} \sin ny \right) \sin mx.$$

For each fixed y $(0 \leqq y \leqq \pi)$ equation (6) is the Fourier sine series representation of the function $f(x,y)$ of the variable x $(0 \leqq x \leqq \pi)$, provided that the coefficients of $\sin mx$ are those in the Fourier sine series; that is,

$$(7) \qquad \sum_{n=1}^{\infty} A_{mn} \sin ny = \frac{2}{\pi} \int_0^{\pi} f(\xi,y) \sin m\xi \, d\xi.$$

The right-hand member here is a sequence of functions $F_m(y)$, $m = 1, 2, \ldots$, each represented by its Fourier sine series (7) on the interval $(0,\pi)$ if

$$A_{mn} = \frac{2}{\pi} \int_0^{\pi} F_m(\eta) \sin n\eta \, d\eta \qquad (n = 1, 2, \ldots).$$

The coefficients therefore have the values

$$(8) \qquad A_{mn} = \frac{4}{\pi^2} \int_0^\pi \int_0^\pi f(\xi,\eta) \sin m\xi \sin n\eta \, d\xi \, d\eta.$$

A formal solution of our membrane problem is given by equation (4) with the coefficients defined by formula (8).

Since the numbers $\sqrt{m^2 + n^2}$ do not change by integral multiples of some fixed number as m and n vary through integral values, the cosine functions in formula (4) have no common period in the variable t, so the displacement z is not generally a periodic function of t. Consequently the vibrating membrane, in contrast to the vibrating string, generally does not produce a musical note. It can be made to do so, however, by giving it the proper initial displacement. If, for example,

$$z(x,y,0) = A \sin x \sin y,$$

the displacements (4) are given by a single term

$$z(x,y,t) = A \sin x \sin y \cos (a \sqrt{2} \, t);$$

then z is periodic in t with period $\pi \sqrt{2}/a$.

PROBLEMS

✓ **1.** One edge of a square plate with insulated faces is kept at a uniform temperature u_0, and the other three edges are kept at temperature zero. Without solving a boundary value problem, but by superposition of solutions of like problems to obtain the trivial case in which all four edges are at temperature u_0, show why the steady temperature at the center of the given plate must be $u_0/4$.

✓ **2.** A square plate has its faces and its edges $x = 0$ and $x = \pi$ $(0 < y < \pi)$ insulated. Its edges $y = 0$ and $y = \pi$ are kept at temperatures zero and $f(x)$, respectively. Derive this formula for its steady temperatures

$$u(x,y) = \frac{1}{2\pi} a_0 y + \sum_{n=1}^\infty a_n \frac{\sinh ny}{\sinh n\pi} \cos nx,$$

where $\qquad a_n = \frac{2}{\pi} \int_0^\pi f(x) \cos nx \, dx \qquad (n = 0, 1, 2, \ldots).$

Note the result when f is a constant, $f(x) = u_0$.

3. Find the harmonic function $u(x,y)$ in the semi-infinite strip

$0 < x < \pi, y > 0$ such that

$$u(0,y) = u(\pi,y) = 0 \qquad\qquad (y > 0),$$
$$u(x,0) = 1 \qquad\qquad (0 < x < \pi),$$

and $|u(x,y)| < M$, where M is some constant. (This boundedness requirement serves as a condition at the missing upper boundary of the strip.)

$$Ans. \quad u = \frac{4}{\pi} \sum_{n=1}^{\infty} \exp\left[-(2n-1)y\right]\frac{\sin(2n-1)x}{2n-1}.$$

4. Given that the imaginary coefficient of $\log(a + ib)$ is $\arctan(b/a)$ when a and b are real numbers and that

$$(a) \qquad\qquad \log\frac{1+z}{1-z} = 2\sum_{n=1}^{\infty} \frac{1}{2n-1} z^{2n-1} \qquad\qquad (|z| < 1),$$

write $z = e^{-y}e^{ix}$ $(y > 0)$ and equate imaginary components in expansion (a) to indicate that the answer to Problem 3 can be written

$$u(x,y) = \frac{2}{\pi}\arctan\frac{\sin x}{\sinh y} \qquad (0 < x < \pi, y > 0).$$

Verify the answer in that form.

5. Derive the formula for the steady temperatures $u(x,y)$ in a semi-infinite wall $0 < x < c, y > 0$ if the plane faces $x = 0$ and $x = c$ are insulated and $u(x,0) = f(x)$, where f and f' are sectionally continuous $(0 < x < c)$ and if $u(x,y)$ is to be bounded.

6. All four faces of an infinitely long rectangular prism bounded by the planes $x = 0$, $x = b$, $y = 0$, and $y = c$ are kept at temperature zero. If the initial temperature distribution is $f(x,y)$, derive this formula for the temperatures $u(x,y,t)$ in the prism:

$$u = \sum_{m=1}^{\infty} \sum_{n=1}^{\infty} B_{mn} \exp\left[-\pi^2 kt\left(\frac{m^2}{b^2} + \frac{n^2}{c^2}\right)\right] \sin\frac{m\pi x}{b}\sin\frac{n\pi y}{c},$$

where $\qquad B_{mn} = \frac{4}{bc}\int_0^c \sin\frac{n\pi y}{c}\int_0^b f(x,y)\sin\frac{m\pi x}{b}\,dx\,dy.$

7. In case $f(x,y) = g(x)h(y)$ in Problem 6, show that the double series for u reduces to a product of two series,

$$u(x,y,t) = v(x,t)w(y,t),$$

and note that v and w represent temperatures in the slabs $0 \leq x \leq b$ and $0 \leq y \leq c$ with their faces at temperature zero and with initial temperatures $g(x)$ and $h(y)$, respectively.

8. If $v(x,t)$ and $w(y,t)$ satisfy the heat equation for one-dimensional flow, $v_t = kv_{xx}$, $w_t = kw_{yy}$, show by differentiation that their product $u = vw$ satisfies the heat equation $u_t = k(u_{xx} + u_{yy})$. Use this fact to arrive at the solution of Problem 7.

65. An Application of Fourier Integrals.   The face $x = 0$ of a semi-infinite solid $x \geqq 0$ is kept at temperature zero (Fig. 19). Let us find the temperatures $u(x,t)$ in the solid when the initial temperature distribution is $f(x)$, assuming at present that f and f' are sectionally continuous on each finite interval and that f is bounded and absolutely integrable over the positive x axis.

If the solid is considered as a limiting case of a slab $0 \leq x \leq c$ as c increases, some condition corresponding to a thermal condition on the face $x = c$ seems to be needed; otherwise the temperature of that face may be increased in any manner as c

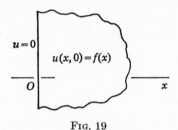

Fig. 19

increases.   We require that our function u be bounded, a condition that also implies that there is no instantaneous source of heat on the face $x = 0$ at the instant $t = 0$.   Then

$$(1) \qquad u_t(x,t) = ku_{xx}(x,t) \qquad\qquad (x > 0, t > 0),$$

$$(2) \qquad \begin{aligned} u(0,t) &= 0 & (t > 0), \\ |u(x,t)| &< M & (x > 0, t > 0), \end{aligned}$$

where M is some constant, and

$$(3) \qquad\qquad u(x,0) = f(x) \qquad\qquad (x > 0).$$

Linear combinations of functions XT will not ordinarily be bounded unless X and T themselves are bounded. Upon separating variables we then have the conditions

$$(4) \qquad \begin{aligned} X''(x) + \lambda X(x) &= 0 & (x > 0), \\ X(0) = 0, \qquad |X(x)| &< M_1, \\ T'(t) + \lambda kT(t) = 0, \qquad |T(t)| &< M_2 & (t > 0), \end{aligned}$$

where M_1 and M_2 are constants.   As pointed out in Sec. 52, the singular eigenvalue problem (4) has continuous eigenvalues $\lambda = \alpha^2$, where α represents *all real positive numbers;* sin αx are the eigenfunctions.   In   this   case   the   corresponding   functions $T = \exp(-\alpha^2 kt)$ are bounded.   The generalized linear combination of the functions XT for all positive α,

$$(5) \qquad u(x,t) = \int_0^\infty g(\alpha) \exp(-\alpha^2 kt) \sin \alpha x \, d\alpha,$$

may satisfy all conditions of the boundary value problem if the function g can be determined so that

$$(6) \qquad f(x) = \int_0^\infty g(\alpha) \sin \alpha x \, d\alpha \qquad\qquad (x > 0).$$

The representation (6) is the Fourier sine integral formula (2), Sec. 52, for our function f if

$$(7) \qquad g(\alpha) = \frac{2}{\pi} \int_0^\infty f(\xi) \sin \alpha \xi \, d\xi \qquad\qquad (\alpha > 0).$$

Our formal solution is therefore

$$(8) \quad u(x,t) = \frac{2}{\pi} \int_0^\infty \exp(-\alpha^2 kt) \sin \alpha x \int_0^\infty f(\xi) \sin \alpha \xi \, d\xi \, d\alpha.$$

We can simplify this result by formally interchanging the order of   integration,   replacing $2 \sin \alpha x \sin \alpha \xi$ by $\cos \alpha(x - \xi) - \cos \alpha(x + \xi)$, and then applying the integration formula (Problem 12, Sec. 66)

$$(9) \qquad \int_0^\infty \exp(-\alpha^2 b) \cos \alpha r \, d\alpha = \frac{1}{2} \sqrt{\frac{\pi}{b}} \exp\left(-\frac{r^2}{4b}\right) \quad (b > 0).$$

Formula (8) then becomes

$$(10) \quad u(x,t) = \frac{1}{2\sqrt{\pi kt}} \int_0^\infty f(\xi) \left\{\exp\left[-\frac{(x-\xi)^2}{4kt}\right] \right.$$
$$\left. - \exp\left[-\frac{(x+\xi)^2}{4kt}\right]\right\} d\xi$$

when $t > 0$.   An alternate form of equation (10), obtained by introducing new variables of integration, is

$$(11) \quad u(x,t) = \frac{1}{\sqrt{\pi}} \int_{-x/(2\sqrt{kt})}^{\infty} \exp{(-\eta^2)} f(x + 2\eta \sqrt{kt}) \, d\eta$$
$$- \frac{1}{\sqrt{\pi}} \int_{x/(2\sqrt{kt})}^{\infty} \exp{(-\eta^2)} f(-x + 2\eta \sqrt{kt}) \, d\eta.$$

The function u defined by formulas (10) and (11) can be established as a solution of our problem under more relaxed conditions on the function f. Assume only that f is sectionally continuous on some interval $(0, x_0)$ and continuous and bounded when $x \geqq x_0$. Then it can be shown from formula (10) that u satisfies the heat equation (1) because the functions

$$t^{-\frac{1}{2}} \exp{[-(x \pm \xi)^2/(4kt)]}$$

satisfy that equation and from formula (11) that u satisfies conditions (2). From the two forms (10) and (11) it can be found that $u(x,t) - f(x) \to 0$ as $t \to 0$ at each point x where f is continuous. Details of the proof are tedious.[1]

When $f(x) = 1$, it follows from formula (11) that

$$(12) \quad u(x,t) = \frac{1}{\sqrt{\pi}} \left[\int_{-x/(2\sqrt{kt})}^{\infty} \exp{(-\eta^2)} \, d\eta \right.$$
$$\left. - \int_{x/(2\sqrt{kt})}^{\infty} \exp{(-\eta^2)} \, d\eta \right].$$

In terms of the *error function*

$$(13) \quad \text{erf}\,(r) = \frac{2}{\sqrt{\pi}} \int_0^r \exp{(-\eta^2)} \, d\eta,$$

where $\text{erf}\,(\infty) = 1$, formula (12) can be written

$$(14) \quad u(x,t) = \text{erf}\left(\frac{x}{2\sqrt{kt}}\right).$$

The full verification of this result is not difficult.

66. Temperatures $u(x,t)$ in an Unlimited Medium. As an application of the general Fourier integral formula we shall derive formulas for the temperatures $u(x,t)$ in a medium that occupies all space, when the initial temperature distribution is

---

[1] A similar verification is carried out on pp. 35ff. of Carslaw and Jaeger's book "Conduction of Heat in Solids," 1947.

$f(x)$. We assume that f is bounded and, for the present, that it satisfies conditions under which it is represented by its Fourier integral formula. The boundary value problem consists of a boundedness condition $|u(x,t)| < M$ and the conditions

(1) $$u_t(x,t) = ku_{xx}(x,t) \quad (-\infty < x < \infty, t > 0),$$
(2) $$u(x,0) = f(x) \qquad (-\infty < x < \infty).$$

Separation of variables leads to the singular eigenvalue problem

$$X''(x) + \lambda X(x) = 0, \qquad |X(x)| < M_1 \quad (-\infty < x < \infty),$$

whose eigenvalues are $\lambda = \alpha^2$, where α is real, and to two linearly independent eigenfunctions $\cos \alpha x$ and $\sin \alpha x$ corresponding to each nonzero value of α. Negative values of α produce no additional eigenfunctions, so we use only the values $\alpha \geqq 0$.

Our generalized linear combination of functions XT becomes

(3) $$u(x,t) = \int_0^\infty \exp(-\alpha^2 kt)[A(\alpha) \cos \alpha x + B(\alpha) \sin \alpha x]\, d\alpha.$$

The coefficients A and B are to be determined so that, when $t = 0$, the integral here represents $f(x)$ $(-\infty < x < \infty)$. According to equations (7) and (8) of Sec. 50 and the Fourier integral theorem (Sec. 51), the representation is valid if

$$A(\alpha) = \frac{1}{\pi} \int_{-\infty}^\infty f(\xi) \cos \alpha\xi \, d\xi, \qquad B(\alpha) = \frac{1}{\pi} \int_{-\infty}^\infty f(\xi) \sin \alpha\xi \, d\xi.$$

Therefore formula (3) becomes

(4) $$u(x,t) = \frac{1}{\pi} \int_0^\infty \exp(-\alpha^2 kt) \int_{-\infty}^\infty f(\xi) \cos \alpha(x - \xi) \, d\xi \, d\alpha.$$

If we formally invert the order of integration here, the integration formula (9) of Sec. 65 can be used to write equation (4) in the form

(5) $$u(x,t) = \frac{1}{2\sqrt{\pi kt}} \int_{-\infty}^\infty f(\xi) \exp\left[-\frac{(x - \xi)^2}{4kt}\right] d\xi \quad (t > 0).$$

An alternate form of this formula is

(6) $$u(x,t) = \frac{1}{\sqrt{\pi}} \int_{-\infty}^\infty f(x + 2\eta \sqrt{kt}) \exp(-\eta^2) \, d\eta.$$

Forms (5) and (6) can be established by assuming only that f is sectionally continuous over some bounded interval, $|x| < c$, and continuous and bounded over the rest of the x axis $|x| \geqq c$. When f is an odd function, u becomes the function found in the preceding section for positive values of x.

PROBLEMS

1. Verify that the function $u = $ erf $(\frac{1}{2}x/\sqrt{kt})$ satisfies the heat equation when $x > 0$, $t > 0$ and these conditions:

$$u(+0,t) = 0 \qquad\qquad (t > 0),$$
$$u(x,+0) = 1 \qquad\qquad (x > 0),$$
$$|u(x,t)| < 1 \qquad\qquad (x > 0, t > 0).$$

2. In case $f(x) = 0$ when $0 < x < c$ and $f(x) = 1$ when $x > c$, show that formula (11), Sec. 65, reduces to

$$u(x,t) = \frac{1}{2} \text{ erf } \left(\frac{c + x}{2\sqrt{kt}}\right) - \frac{1}{2} \text{ erf } \left(\frac{c - x}{2\sqrt{kt}}\right).$$

Verify this solution.

3. The face $x = 0$ of a semi-infinite solid is kept at constant temperature u_0 after the solid $x > 0$ is initially at temperature zero throughout. Obtain a formula for the temperatures $u(x,t)$ in the body.

$$Ans. \quad u = u_0 \left[1 - \text{erf }\left(\frac{x}{2\sqrt{kt}}\right)\right].$$

4. The face $x = 0$ of a semi-infinite solid $x \geqq 0$ is insulated. If the initial temperature distribution is $f(x)$, derive the temperature formula

$$u(x,t) = \frac{1}{\sqrt{\pi}} \int_{-x/(2\sqrt{kt})}^{\infty} f(x + 2\eta \sqrt{kt}) \exp(-\eta^2) \, d\eta$$
$$+ \frac{1}{\sqrt{\pi}} \int_{x/(2\sqrt{kt})}^{\infty} f(-x + 2\eta \sqrt{kt}) \exp(-\eta^2) \, d\eta.$$

5. In Problem 4, if $f(x) = 1$ when $0 < x < c$ and $f(x) = 0$ when $x > c$, show that

$$u(x,t) = \frac{1}{2} \text{ erf } \left(\frac{c - x}{2\sqrt{kt}}\right) + \frac{1}{2} \text{ erf } \left(\frac{c + x}{2\sqrt{kt}}\right).$$

6. If the initial temperature distribution $f(x)$ in the unlimited medium (Sec. 66) is $f(x) = 0$ $(x < 0)$ and $f(x) = 1$ $(x > 0)$, show that

$$u(x,t) = \frac{1}{2} + \frac{1}{2} \text{ erf } \left(\frac{x}{2\sqrt{kt}}\right).$$

Verify this solution of the boundary value problem.

7. Derive this solution of the wave equation $y_{tt} = a^2 y_{xx}$, $-\infty < x < \infty$, $t > 0$ that satisfies the conditions $y(x,0) = f(x)$, $y_t(x,0) = 0$, when $-\infty < x < \infty$:

$$y(x,t) = \frac{1}{\pi} \int_0^\infty \cos \alpha at \int_{-\infty}^\infty f(\xi) \cos \alpha(x - \xi)\, d\xi\, d\alpha.$$

Also, reduce the solution to the form obtained in Example 2 of Sec. 20, namely,

$$y(x,t) = \tfrac{1}{2}[f(x + at) + f(x - at)].$$

8. A semi-infinite string with one end fixed at the origin is stretched along the positive half of the x axis and released at rest from a position $y = f(x)$, $x \geq 0$. Derive the formula

$$y(x,t) = \frac{2}{\pi} \int_0^\infty \cos \alpha at \sin \alpha x \int_0^\infty f(\xi) \sin \alpha \xi\, d\xi\, d\alpha$$

for its transverse displacements. If $F_1(x)$, $-\infty < x < \infty$, is the odd extension of f, show that the formula reduces to the form

$$y(x,t) = \tfrac{1}{2}[F_1(x + at) + F_1(x - at)].$$

9. Find the harmonic function $V(x,y)$ in the semi-infinite strip $x > 0$

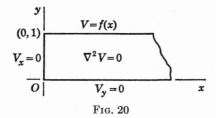

Fig. 20

$0 < y < 1$ that satisfies the boundary conditions (Fig. 20)

$$V_x(0,y) = 0, \qquad V_y(x,0) = 0, \qquad V(x,1) = f(x),$$

where $f(x) = 1$ when $0 < x < 1$ and $f(x) = 0$ when $x > 1$, and where $V(x,y)$ is to be bounded over the strip.

$$Ans. \quad V = \frac{2}{\pi} \int_0^\infty \frac{\sin \alpha \cos \alpha x \cosh \alpha y}{\alpha \cosh \alpha}\, d\alpha.$$

10. If $V(x,y)$ is bounded and harmonic in the quadrant $x > 0$, $y > 0$ and

$$V(0,y) = 0 \qquad\qquad\qquad (y > 0),$$
$$V(x,0) = f(x) \qquad\qquad\quad (x > 0),$$

derive this formula for V when $x \geq 0$ and $y > 0$:

$$V(x,y) = \frac{y}{\pi} \int_0^\infty f(\xi) \left[\frac{1}{y^2 + (\xi - x)^2} - \frac{1}{y^2 + (\xi + x)^2} \right] d\xi.$$

When $f(x) = 1$, show that

$$V(x,y) = \frac{2}{\pi} \arctan \frac{x}{y}.$$

11. When a semi-infinite solid $x \geq 0$ initially at a uniform temperature is cooled or heated by keeping its boundary at a uniform constant temperature (Sec. 65), show that the times required for two interior points to reach the same temperature are proportional to the squares of the distances of the points from the boundary plane.

12. Derive the integration formula (9), Sec. 65, by first writing

$$y(r) = \int_0^\infty \exp(-\alpha^2 b) \cos \alpha r \, d\alpha \qquad (b > 0),$$

then differentiating the integral to find $y'(r)$. Integrate the new integral by parts to show that $2by'(r) = -ry(r)$, note that $y(0) = \frac{1}{2} \sqrt{\pi/b}$, and solve for $y(r)$.

67. Observations and Further Examples.

Not all linear homogeneous partial differential equations yield to the method of separation of variables. Conditions for the existence of particular solutions $V = X(x)Y(y)$ of the equation

$$V_{xx}(x,y) + (x + y)^2 V_{yy}(x,y) = 0,$$

for instance, do not lead to differential equations of the Sturm-Liouville type in either X or Y.

Although the separation process does apply to Laplace's equation in rectangular, cylindrical, or spherical coordinates, to certain wave equations and heat equations involving the laplacian operator in those coordinates, and to many other equations, the general test consists of trying out the process on each equation.

a. Periodic Boundary Conditions. The following example will illustrate problems that involve expansions in the basic Fourier series containing both sine and cosine terms.

Let $u(\rho,\phi)$ denote the steady temperatures in a long solid cylinder $\rho \leq 1$, $-\infty < z < \infty$, when the temperature of the surface $\rho = 1$ is a given function $f(\phi)$, where ρ, ϕ and z are cylindrical coordinates.   Then (Sec. 9)

(1) $\rho^2 u_{\rho\rho}(\rho,\phi) + \rho u_\rho(\rho,\phi) + u_{\phi\phi}(\rho,\phi) = 0$ $(\rho < 1, -\pi < \phi \leqq \pi)$,
(2) $\qquad\qquad\qquad u(1,\phi) = f(\phi)$ $\qquad\qquad (-\pi < \phi < \pi)$.

Also, u and its partial derivatives of the first order are to be continuous interior to the cylinder.

If functions of type $R(\rho)\Phi(\phi)$ are to satisfy equation (1) and the continuity requirements, we find that

(3) $\Phi''(\phi) + \lambda\Phi(\phi) = 0$, $\quad \Phi(-\pi) = \Phi(\pi)$, $\quad \Phi'(-\pi) = \Phi'(\pi)$,
(4) $\qquad\qquad \rho^2 R''(\rho) + \rho R'(\rho) - \lambda R(\rho) = 0 \qquad (0 \leqq \rho < 1)$.

The eigenvalues of problem (3), whose boundary conditions are of periodic type, are the numbers $\lambda = 0$ and $\lambda = n^2$ $(n = 1, 2, \ldots)$, as noted in Sec. 34.   The set $\{1, \cos n\phi, \sin n\phi\}$ is a set of corresponding eigenfunctions, orthogonal on the interval $(-\pi,\pi)$.

Equation (4) is of Cauchy type.   When $\lambda = n^2$, its general solution is $R = C_1\rho^n + C_2\rho^{-n}$; when $\lambda = 0$, $R = C_3 \log \rho + C_4$. If R is to be continuous on the axis $\rho = 0$, then $C_2 = C_3 = 0$. The generalized linear combination of our continuous functions $R\Phi$ can be written

(5) $\qquad u(\rho,\phi) = \tfrac{1}{2}a_0 + \sum_{n=1}^{\infty} \rho^n(a_n \cos n\phi + b_n \sin n\phi)$.

This satisfies boundary condition (2) if a_n and b_n, including a_0, are the coefficients in the Fourier series for f on the interval $(-\pi,\pi)$,

$$a_n = \frac{1}{\pi} \int_{-\pi}^{\pi} f(\phi) \cos n\phi \, d\phi, \qquad b_n = \frac{1}{\pi} \int_{-\pi}^{\pi} f(\phi) \sin n\phi \, d\phi.$$

We assume that f satisfies the conditions in our Fourier theorem (Sec. 41).

b. *Sturm-Liouville Series.*   Even when separation of variables applies, the superposition process may lead to a generalized Fourier series representation of a given function.   Our formal procedure may then include an assumption that the representation is valid.

As an example, let $u(x,t)$ denote temperatures in a slab bounded by the planes $x = 0$ and $x = 1$, initially at temperatures $f(x)$, when $u = 0$ on the face $x = 0$ and surface heat transfer takes place at the face $x = 1$ into a medium at temperature zero at a rate per unit area proportional to $[u(1,t) - 0]$ (Fig. 21). If $k = 1$ and h is a positive constant, a coefficient of surface heat transfer, then

$$(6) \qquad\qquad u_t(x,t) = u_{xx}(x,t) \qquad (0 < x < 1, t > 0),$$
$$(7) \quad u(0,t) = 0, \qquad u_x(1,t) = -hu(1,t), \qquad u(x,0) = f(x).$$

Here $T'(t) = -\lambda T(t)$, and the Sturm-Liouville problem is

$$X''(x) + \lambda X(x) = 0, \qquad X(0) = 0, \qquad hX(1) + X'(1) = 0.$$

The full set of eigenvalues, found in Problem 6, Sec. 34, is

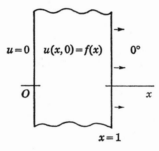

FIG. 21

$\lambda = \alpha_n{}^2$, where α_n are the positive roots of the equation $\tan \alpha = -\alpha/h$. The normalized eigenfunctions are found to be

$$(8) \qquad\qquad \phi_n(x) = \left(\frac{2h}{h + \cos^2 \alpha_n}\right)^{\frac{1}{2}} \sin \alpha_n x \quad (n = 1, 2, \ldots),$$

and the formal solution of our problem is

$$(9) \qquad\qquad u(x,t) = \sum_{n=1}^{\infty} c_n \exp\left(-\alpha_n{}^2 t\right) \phi_n(x),$$

where, in order that $u(x,0) = f(x)$ when $0 < x < 1$,

$$(10) \quad c_n = \int_0^1 f(x)\phi_n(x)\, dx = \left(\frac{2h}{h + \cos^2 \alpha_n}\right)^{\frac{1}{2}} \int_0^1 f(x) \sin \alpha_n x\, dx.$$

We have *assumed* that our function f is represented by its generalized Fourier series with respect to the orthonormal eigenfunctions (8)

$$f(x) = \sum_{n=1}^{\infty} c_n \phi_n(x) \quad [0 < x < 1, c_n = (f, \phi_n)].$$

Representation theorems in the Sturm-Liouville theory, proved in references cited in Sec. 32, show that the series does converge to $f(x)$ at each point x $(0 < x < 1)$ where f is continuous, provided that f and f' are sectionally continuous on the interval $(0,1)$.

PROBLEMS

1. When $f(x) = 1$ $(0 < x < 1)$ in the boundary value problem consisting of equations (6) and (7), Sec. 67, show that

$$u(x,t) = 2h \sum_{n=1}^{\infty} \frac{1 - \cos \alpha_n}{\alpha_n(h + \cos^2 \alpha_n)} \exp(-\alpha_n^2 t) \sin \alpha_n x.$$

2. Find the harmonic function $u(\rho, \phi)$ in the quadrant $\rho < 1$, $0 < \phi < \frac{1}{2}\pi$ of a unit circle in the plane $z = 0$ if $u = 0$ on the radii $\phi = 0$ and $\phi = \frac{1}{2}\pi$ and $u = f(\phi)$ on the arc $\rho = 1$, $0 < \phi < \frac{1}{2}\pi$.

$$Ans. \quad u = \frac{4}{\pi} \sum_{n=1}^{\infty} \rho^{2n} \sin 2n\phi \int_0^{\frac{1}{2}\pi} f(\theta) \sin 2n\theta \, d\theta.$$

3. A string stretched between the points $(0,0)$ and $(\pi,0)$ is released at rest from an initial displacement $y = f(x)$. Its motion is opposed by air resistance proportional to the velocity at each point. Let the unit of time be chosen so that the equation of motion becomes

$$y_{tt}(x,t) = y_{xx}(x,t) - 2hy_t(x,t).$$

When $0 < h < 1$ (h constant), derive the formula

$$y(x,t) = e^{-ht} \sum_{n=1}^{\infty} b_n \left(\cos K_n t + \frac{h}{K_n} \sin K_n t \right) \sin nx$$

for the transverse displacements, where

$$K_n = (n^2 - h^2)^{\frac{1}{2}}, \qquad b_n = \frac{2}{\pi} \int_0^{\pi} f(x) \sin nx \, dx.$$

4. Find the harmonic function $V(x,y)$ in the strip $0 < y < b$, $-\infty < x < \infty$ such that $V(x,0) = 0$ and $V(x,b) = f(x)$ $(-\infty < x < \infty)$, where f is bounded and represented by its Fourier integral formula and $V(x,y)$ is to be bounded in the strip.

$$\text{Ans.}\quad V(x,y) = \frac{1}{\pi} \int_0^\infty \frac{\sinh \alpha y}{\sinh \alpha b} \int_{-\infty}^\infty f(\xi) \cos \alpha(x - \xi) \, d\xi \, d\alpha.$$

5. A bounded harmonic function $u(x,y)$ in the semi-infinite strip $x > 0, 0 < y < 1$ is to satisfy the boundary conditions

$$u(0,y) = 1 \qquad\qquad (0 < y < 1);$$
$$u_y(x,0) = 0, \qquad u_y(x,1) = -u(x,1) \qquad (x > 0).$$

Assuming the validity of the Sturm-Liouville expansion that arises, derive the formula

$$u(x,y) = 2 \sum_{n=1}^\infty \frac{\sin \alpha_n}{\alpha_n(1 + \sin^2 \alpha_n)} \exp(-\alpha_n x) \cos \alpha_n y,$$

where α_n are the positive roots of the equation $\tan \alpha = 1/\alpha$.

6. Solve the following boundary value problem for steady temperatures $u(x,y)$ in a thin plate in the shape of a semi-infinite strip when heat transfer to the surroundings at temperature zero takes place at the faces of the plate.

$$u_{xx}(x,y) + u_{yy}(x,y) - hu(x,y) = 0 \quad (x > 0, 0 < y < 1),$$
$$u_x(0,y) = 0 \qquad\qquad (0 < y < 1);$$
$$u(x,0) = 0, \qquad u(x,1) = f(x) \qquad\qquad (x > 0),$$

where h is a positive constant and $f(x) = 1$ when $0 < x < c$, $f(x) = 0$ when $x > c$.

$$\text{Ans.}\quad u(x,y) = \frac{2}{\pi} \int_0^\infty \frac{\sin \alpha c \, \cos \alpha x \, \sinh (y \sqrt{\alpha^2 + h})}{\alpha \sinh \sqrt{\alpha^2 + h}} \, d\alpha.$$

7. Let $u(r,t)$ denote *temperatures in a solid sphere* bounded by the surface $r = c$, where r is the spherical coordinate and when the solid is initially at a uniform temperature u_0 and the surface is then kept at temperature zero.   Then

$$\frac{\partial u}{\partial t} = \frac{k}{r} \frac{\partial^2}{\partial r^2}(ru), \qquad u(c,t) = 0, \qquad u(r,0) = u_0 \quad (0 \le r < c).$$

Introduce the new unknown function $v(r,t) = ru(r,t)$ and note that $v(0,t) = 0$ because u is to be continuous at the center $r = 0$.   Thus derive the formula

$$u(r,t) = \frac{2u_0 c}{\pi} \sum_{n=1}^\infty (-1)^{n+1} \exp\left(-\frac{n^2\pi^2 kt}{c^2}\right) \frac{1}{nr} \sin \frac{n\pi r}{c}.$$

8. A spherical body 40 cm in diameter, initially at 100°C throughout, is cooled by keeping its surface at 0°C (see Problem 7). Find the approximate temperature of its center 10 min after cooling begins if the material is (*a*) iron, for which $k = 0.15$ cgs unit; (*b*) concrete, for which $k = 0.005$ cgs unit.   *Ans.*   (*a*) 22°C; (*b*) 100°C.

9. The boundary $r = 1$ of a solid sphere is kept insulated. Initially the solid is at temperatures $f(r)$. Assuming the validity of the Sturm-Liouville expansion that arises (Problem 5, Sec. 34), derive this formula for the temperatures in the sphere:

$$u(r,t) = 3 \int_0^1 \xi^2 f(\xi) \, d\xi + \sum_{n=1}^{\infty} A_n \exp\left(-\alpha_n^2 kt\right) \frac{\sin \alpha_n r}{r},$$

where α_n are the positive roots of the equation $\tan \alpha = \alpha$ and

$$A_n = \frac{2}{\sin^2 \alpha_n} \int_0^1 \xi f(\xi) \sin \alpha_n \xi \, d\xi.$$

10. For any constant C, verify that the function

$$v(x,t) = Cxt^{-\frac{3}{2}} \exp\left[-x^2/(4kt)\right]$$

satisfies the heat equation when $x > 0$ and $t > 0$. Also verify that $v(+0,t) = 0$ when $t > 0$ and that $v(x,+0) = 0$ when $x > 0$. Thus v can be added to the function u found in Sec. 65 to form other solutions of the problem there if the temperature function were not required to be bounded. But note that v is unbounded as x and t tend to zero, as can be seen by letting x vanish while $t = x^2$.

CHAPTER 8

BESSEL FUNCTIONS AND APPLICATIONS

68. Bessel's Equation. In boundary value problems that involve the laplacian $\nabla^2 u$ expressed in cylindrical coordinates, the process of separating variables often produces an equation of the form

$$(1) \qquad \rho^2 \frac{d^2 y}{d\rho^2} + \rho \frac{dy}{d\rho} + (\lambda^2 \rho^2 - \nu^2) y = 0$$

in the function y of the cylindrical coordinate ρ. In such problems, as we shall observe in examples presented in this chapter, the parameter λ^2 is the separation constant whose values are the eigenvalues associated with equation (1). The parameter ν is a real number determined by other aspects of the boundary value problem; it is most commonly either zero or a positive integer.

When written in terms of the variable x, where

$$x = \lambda \rho,$$

equation (1) takes a form free of the parameter λ:

$$(2) \qquad x^2 y''(x) + x y'(x) + (x^2 - \nu^2) y(x) = 0.$$

This linear homogeneous differential equation is *Bessel's equation*. Its solutions are called *Bessel functions* or, sometimes, cylindrical functions.

Upon comparing equation (2) with the standard form

$$y''(x) + A(x) y'(x) + B(x) y(x) = 0,$$

we see that $A(x) = 1/x$ and $B(x) = 1 - \nu^2/x^2$. Those coefficients are continuous except at the origin, which is a singular point of Bessel's equation. The existence and uniqueness theorem stated in Sec. 19 applies to the equation for an interval that does not include the origin. But for boundary value problems in regions interior to circles or cylinders $\rho = c$, the origin $x = 0$, corresponding to the center or axis $\rho = 0$, is interior to the region.   The interval for the variable x then has the origin as an end point.

We limit our attention primarily to the cases $\nu = n$, where $n = 0, 1, 2, \ldots$.   In those cases we shall see that there is a solution of Bessel's equation which is *analytic for all values of x,* including the origin; that is, the solution is represented by a power series in x, convergent for every x.   That solution, denoted by $J_n(x)$, and its derivatives of all orders are therefore everywhere continuous functions.

69. Bessel Functions J_n.   We seek a solution of Bessel's equation

$$(1) \quad x^2 y''(x) + x y'(x) + (x^2 - n^2) y(x) = 0 \quad (n = 0, 1, 2, \ldots)$$

in the form of a power series multiplied by x^p, where p is some constant.   That is, we attempt to determine p and the coefficients a_j so that the function

$$(2) \qquad\qquad y = x^p \sum_{j=0}^{\infty} a_j x^j = \sum_{j=0}^{\infty} a_j x^{p+j}$$

satisfies equation (1).   Here a_0 represents the coefficient of the first nonvanishing term in the series, so that $a_0 \neq 0$.

Assume for the present that the series is differentiable.   Then upon substituting the function (2) and its derivatives into equation (1) we obtain the equation

$$\sum_{j=0}^{\infty} [(p+j)(p+j-1) + (p+j) + (x^2 - n^2)] a_j x^{p+j} = 0,$$

or $\qquad \displaystyle\sum_{j=0}^{\infty} [(p+j)^2 - n^2] a_j x^j + \sum_{k=0}^{\infty} a_k x^{k+2} = 0.$

The last sum can be written

$$\sum_{j=2}^{\infty} a_{j-2} x^j;$$

thus the equation becomes

$$(3) \quad (p-n)(p+n)a_0 + (p-n+1)(p+n+1)a_1 x$$

$$+ \sum_{j=2}^{\infty} [(p-n+j)(p+n+j)a_j + a_{j-2}] x^j = 0.$$

Equation (3) is an identity in x if the coefficient of each power of x vanishes. This condition is satisfied if $p = n$ or $p = -n$ so that the constant term vanishes and if $a_1 = 0$ and

$$(p - n + j)(p + n + j)a_j + a_{j-2} = 0 \quad (j = 2, 3, \ldots).$$

We make the choice $p = n$; then the *recurrence relation*

$$(4) \qquad a_j = -\frac{1}{j(2n + j)} a_{j-2} \qquad (j = 2, 3, \cdots)$$

is obtained, giving each coefficient in terms of the second one preceding it in the series. Note that $2n + j \neq 0$ in equation (4). The choice $p = -n$ fails to give a recurrence relation.

Since $a_1 = 0$, relation (4) requires that $a_3 = 0$, then that $a_5 = 0$, etc.; thus

$$(5) \qquad a_{2j-1} = 0 \qquad (j = 1, 2, \ldots).$$

For the remaining coefficients the recurrence relation is

$$(6) \qquad a_{2j} = \frac{-1}{2^2 j(n + j)} a_{2j-2} \qquad (j = 1, 2, \ldots).$$

Therefore $\quad a_{2j-2} = \dfrac{-1}{2^2(j - 1)(n + j - 1)} a_{2j-4}$

and so $\quad a_{2j} = \dfrac{(-1)^2}{j(j - 1)} \dfrac{1}{(n + j)(n + j - 1)} \dfrac{1}{2^4} a_{2j-4}.$

Continuing this process through $j - 2$ further steps, we find that

$$a_{2j} = \frac{(-1)^j}{j!} \frac{1}{(n + j)(n + j - 1) \cdots (n + 1)} \frac{a_0}{2^{2j}}.$$

Thus a_0 is a common factor in all terms of the series. To simplify the series we select this value for a_0:

$$a_0 = \frac{1}{2^n n!}.$$

Then our formula for the nonvanishing coefficients becomes

$$(7) \qquad a_{2j} = \frac{(-1)^j}{j!(n + j)!} \frac{1}{2^{n+2j}} \quad (j = 0, 1, 2, \ldots),$$

where we use the convention that $0! = 1$.

Our proposed solution (2) of Bessel's equation can now be

written, in view of formulas (5) and (7),

$$y = J_n(x),$$

where J_n is *Bessel's function of the first kind of order n*, or *index n*, defined by the equation

$$(8) \quad J_n(x) = \sum_{j=0}^{\infty} \frac{(-1)^j}{j!(n+j)!} \left(\frac{x}{2}\right)^{n+2j} \qquad (n = 0, 1, 2, \ldots)$$

$$= \frac{x^n}{2^n n!} \left[1 - \frac{x^2}{2^2(n+1)} + \frac{x^4}{2^4 2!(n+1)(n+2)} - \cdots \right].$$

The power series (8) is absolutely convergent for all x $(-\infty < x < \infty)$, according to the ratio test. Hence it is differentiable with respect to x, any number of times. Since its coefficients satisfy the recurrence relation needed to make its sum satisfy Bessel's equation when the series is differentiable, then $y = J_n(x)$ is a solution of that equation.

Theorem 1. *For all x, the analytic function $J_n(x)$ is a particular solution of Bessel's equation (1).*

Since Bessel's equation is homogeneous, the function $CJ_n(x)$, where C is any constant, is also a solution.

From formula (8) we note that

$$(9) \qquad\qquad J_n(-x) = (-1)^n J_n(x);$$

that is, J_n is an even function if $n = 0, 2, 4, \ldots$, but odd if $n = 1, 3, 5, \ldots$.

Note that the series representation of J_0,

$$(10) \qquad J_0(x) = 1 - \frac{x^2}{2^2} + \frac{x^4}{2^2 4^2} - \frac{x^6}{2^2 4^2 6^2} + \cdots,$$

bears some resemblance to the power series for $\cos x$. There is also a similarity between the power series representations of the odd functions $J_1(x)$ and $\sin x$. Similarities between the properties of those functions include, as we shall see, the differentiation formula $J_0'(x) = -J_1(x)$, corresponding to the formula for the derivative of $\cos x$. Graphs of J_0 and J_1 will be shown in Sec. 75.

70. Some Other Bessel Functions.   Functions linearly independent of J_n that satisfy Bessel's equation

(1) $x^2y''(x) + xy'(x) + (x^2 - n^2)y(x) = 0$ $(n = 0, 1, 2, \ldots)$

can be obtained by various methods of fairly elementary nature.

The singular point $x = 0$ of equation (1) belongs to the class known as regular singular points. The power series procedure, extended so as to give general solutions near regular singular points, applies to Bessel's equation.[1]   It gives a solution $y = Y_n(x)$ where Y_n, *Bessel's function of the second kind*, is represented by the sum of $J_n(x)$ log x and a power series that converges for all x. In particular, when $x > 0$, $Y_0(x)$ is a certain linear combination of $J_0(x)$ and the function

(2) $J_0(x) \log x + \dfrac{x^2}{2^2} - \dfrac{x^4}{2^2 4^2}\left(1 + \dfrac{1}{2}\right)$

$$+ \dfrac{x^6}{2^2 4^2 6^2}\left(1 + \dfrac{1}{2} + \dfrac{1}{3}\right) - \cdots$$

Since Y_n is unbounded and J_n is bounded as $x \to 0$, it is clear that Y_n is not a constant times J_n; that is, Y_n and J_n are linearly independent solutions of Bessel's equation.   If A and B are arbitrary constants, the general solution of that equation is

(3) $y = AJ_n(x) + BY_n(x)$ $(n = 0, 1, 2, \ldots ; x > 0)$.

An integration by parts shows that the *gamma function*

(4) $\Gamma(\nu) = \displaystyle\int_0^\infty e^{-t}t^{\nu-1}\, dt$ $(\nu > 0)$

has the factorial property

(5) $\Gamma(\nu + 1) = \nu\Gamma(\nu)$

when $\nu > 0$. That property is *assigned* to the function when $\nu < 0$, so that $\Gamma(\nu) = \Gamma(\nu + 1)/\nu$ when $-1 < \nu < 0$ or when $-2 < \nu < -1$, etc.; thus conditions (4) and (5) together define $\Gamma(\nu)$ for all ν except $\nu = 0, -1, -2, \ldots$ We find from equation (4) that $\Gamma(1) = 1$; also it can be shown that Γ is continuous

---

[1] The procedure is explained in E. D. Rainville, "Elementary Differential Equations," 2d ed., chap. 19, 1958. The method depends on the nature of the roots of the indicial equation, the condition on the exponent p. For Bessel's equation the indicial equation is $p^2 - n^2 = 0$, as seen from equation (3), Sec. 69.

when $\nu > 0$.   Then it follows from property (5) that $\Gamma(+0) = \infty$ and consequently that $|\Gamma(\nu)|$ becomes infinite as $\nu \to -n$ ($n = 1, 2, \ldots$).

When $\nu = 2, 3, \ldots$, $\Gamma(\nu)$ reduces to a factorial; specifically,

$$(6) \qquad \Gamma(n + 1) = n! \qquad (n = 1, 2, \ldots).$$

The proof of property (6) and the further property that

$$\Gamma(\tfrac{1}{2}) = \sqrt{\pi}$$

is left to the problems.

Now consider Bessel's equation

$$(7) \qquad x^2 y''(x) + x y'(x) + (x^2 - \nu^2) y(x) = 0$$

in which ν is any real number.   The procedure used in Sec. 69 can be modified, by using the factorial property of the gamma function, to derive the solution

$$y = J_\nu(x),$$

where J_ν is *Bessel's function of the first kind, of index ν*:

$$(8) \qquad J_\nu(x) = \sum_{j=0}^{\infty} \frac{(-1)^j}{j!\,\Gamma(\nu + j + 1)} \left(\frac{x}{2}\right)^{\nu + 2j}.$$

In case ν is a negative integer those terms for which the argument $(\nu + j + 1)$ of Γ has values zero or a negative integer are to be dropped from the series.   To verify that J_ν satisfies equation (7) is not difficult.   Note that our definition (8), Sec. 69, of J_n is a special case of the above definition of J_ν.

The series (8) is a product of x^ν by a power series in x that converges for all x.   Thus when $\nu \neq \pm n$, where $n = 0, 1, 2, \ldots$, either J_ν or some of its derivatives fail to exist when $x = 0$. When ν has none of the values $\pm n$, either J_ν or $J_{-\nu}$ is unbounded as $x \to 0$ and the other tends to zero as $x \to 0$.   Therefore J_ν and $J_{-\nu}$ are linearly independent functions, and the general solution of Bessel's equation (7) can be written

$$(9) \qquad y = A J_\nu(x) + B J_{-\nu}(x) \qquad (\nu \neq 0, \pm 1, \pm 2, \ldots).$$

It is left to the problems to show that J_n and J_{-n} are linearly dependent because

$$(10) \qquad J_{-n}(x) = (-1)^n J_n(x) \qquad (n = 1, 2, \ldots).$$

71. Differentiation and Recurrence Formulas.   Since

$$x^{-n}J_n(x) = \frac{1}{2^n} \sum_{j=0}^{\infty} \frac{(-1)^j}{j!(n+j)!} \left(\frac{x}{2}\right)^{2j} \qquad (n = 0, 1, 2, \ldots),$$

it follows that

$$\frac{d}{dx}[x^{-n}J_n(x)] = \frac{1}{2^n} \sum_{j=1}^{\infty} \frac{j(-1)^j}{j(j-1)!(n+j)!} \left(\frac{x}{2}\right)^{2j-1}$$

$$= x^{-n} \left(\frac{x}{2}\right)^n \sum_{k=0}^{\infty} \frac{(-1)^{k+1}}{k!(n+k+1)!} \left(\frac{x}{2}\right)^{2k+1}$$

$$= -x^{-n} \sum_{k=0}^{\infty} \frac{(-1)^k}{k!(n+1+k)!} \left(\frac{x}{2}\right)^{n+1+2k}.$$

That is,

(1) $$\frac{d}{dx}[x^{-n}J_n(x)] = -x^{-n}J_{n+1}(x) \qquad (n = 0, 1, 2, \ldots).$$

As a special case we have the formula

(2) $$J_0'(x) = -J_1(x).$$

Similarly, from the power series representation of $x^n J_n(x)$ we can show that

(3) $$\frac{d}{dx}[x^n J_n(x)] = x^n J_{n-1}(x) \qquad (n = 1, 2, \ldots).$$

Formulas (1) and (3) can be written

$$xJ_n'(x) - nJ_n(x) = -xJ_{n+1}(x),$$
$$xJ_n'(x) + nJ_n(x) = xJ_{n-1}(x),$$

and upon eliminating $J_n'(x)$ from these equations we find that

(4) $$xJ_{n+1}(x) = 2nJ_n(x) - xJ_{n-1}(x) \quad (n = 1, 2, \ldots).$$

This *recurrence formula* expresses J_{n+1} in terms of the functions J_n and J_{n-1} with lower indices.

From formula (3) we get the integration formula

(5)
$$\int_0^r x^n J_{n-1}(x)\, dx = r^n J_n(r) \quad (n = 1, 2, \ldots).$$

An important special case is

(6)
$$\int_0^r x J_0(x)\, dx = r J_1(r).$$

Formulas (1), (3), and (4) are valid when n is replaced by the unrestricted parameter ν. Modifications of the derivations simply consist of writing $\Gamma(\nu + j + 1)$ or $(\nu + j)\Gamma(\nu + j)$ in place of $(n + j)!$.

PROBLEMS

1. Show that

$$J_0(0) = 1, \qquad J_n(0) = 0 \qquad \text{if } n = 1, 2, \ldots.$$

2. Derive formula (3), Sec. 71.

3. Establish the differentiation formula

$$x^2 J_n''(x) = (n^2 - n - x^2) J_n(x) + x J_{n+1}(x) \qquad (n = 0, 1, 2, \ldots).$$

4. From the series representation of J_n show that

$$i^{-n} J_n(ix) = \sum_{j=0}^{\infty} \frac{1}{j!(n + j)!} \left(\frac{x}{2}\right)^{n+2j} \qquad (n = 0, 1, 2, \ldots).$$

The function $I_n(x) = i^{-n} J_n(ix)$ is the *modified Bessel function of the first kind.* Show that the series here converges for all x; also show that $I_n(x) > 0$ when $x > 0$, that $I_n(-x) = (-1)^n I_n(x)$, and that since $J_n(x)$ satisfies Bessel's equation, then I_n satisfies the modified equation

$$t^2 I_n''(t) + t I_n'(t) - (t^2 + n^2) I_n(t) = 0.$$

5. (a) Derive the factorial property $\Gamma(\nu + 1) = \nu \Gamma(\nu)$ of the gamma function where $\nu > 0$. (b) Show that $\Gamma(1) = 1$ and that $\Gamma(n + 1) = n!$ when $n = 1, 2, \ldots$.

6. Show that

$$\Gamma(\tfrac{1}{2}) = 2 \int_0^\infty e^{-x^2}\, dx = \sqrt{\pi}\ \text{erf}\ (\infty) = \sqrt{\pi}.$$

7. From definition (8), Sec. 70, of J_ν show that

(a) $\quad J_{-\frac{1}{2}}(x) = \left(\frac{2}{\pi x}\right)^{\frac{1}{2}} \cos x; \qquad$ (b) $J_{\frac{1}{2}}(x) = \left(\frac{2}{\pi x}\right)^{\frac{1}{2}} \sin x.$

8. Use formula (8), Sec. 70, and recall that terms in which $\nu + j + 1$ is zero or a negative integer are to be dropped from the series to show that

$$J_{-n}(x) = (-1)^n J_n(x) \qquad (n = 1, 2, \ldots);$$

hence that the functions J_n and J_{-n} are linearly dependent.

9. Verify that the function J_ν defined by formula (8), Sec. 70, satisfies Bessel's equation with index ν.

10. Derive the differentiation formula

$$\frac{d}{dx}[x^\nu J_\nu(x)] = x^\nu J_{\nu-1}(x) \qquad (\nu \text{ unrestricted}).$$

72. Integral Forms of J_n. Consider two absolutely convergent power series with sums $\alpha(x)$ and $\beta(x)$:

$$(1) \qquad \sum_{k=0}^{\infty} a_k x^k = \alpha(x), \qquad \sum_{k=0}^{\infty} b_k x^k = \beta(x).$$

Then the *Cauchy product* of those series converges absolutely, and its sum is the product of their sums[1]; that is,

$$(2) \qquad \sum_{m=0}^{\infty} c_m x^m = \alpha(x)\beta(x), \quad \text{where } c_m = \sum_{k=0}^{m} a_k b_{m-k}.$$

In order to associate the coefficients c_j with the coefficients $(-1)^j[j!(n+j)!2^{2j}]^{-1}$ in the power series representing $J_n(x)$ we now let α and β depend also on parameters θ and n $(n = 0, 1, 2, \ldots)$ as follows:

$$(3) \qquad \alpha(x,\theta) = \exp\left(\tfrac{1}{2}ixe^{i\theta}\right), \qquad \beta(x,\theta) = \exp\left(\tfrac{1}{2}ixe^{-i\theta}\right)e^{in\theta}.$$

Then from the representation of $\exp z$ in powers of z, absolutely convergent for all complex z, we see that the coefficients of x^k in series (1) are

$$(4) \qquad a_k(\theta) = \frac{i^k}{2^k k!}e^{ik\theta}, \qquad b_k(\theta) = \frac{i^k}{2^k k!}e^{i(n-k)\theta},$$

and if we write

$$d_k = \frac{i^k}{2^k k!},$$

[1] Kaplan, W., "Advanced Calculus," p. 334, 1952.

the coefficients in the Cauchy product (2) become

$$
(5) \qquad c_m(\theta) = \sum_{k=0}^{m} d_k e^{ik\theta} d_{m-k} e^{i(n-m+k)\theta}
$$

$$
= \sum_{k=0}^{m} d_k d_{m-k} e^{i(2k-m+n)\theta}.
$$

Since
$$
\int_{-\pi}^{\pi} e^{ip\theta}\, d\theta = 0 \qquad \text{if } p = \pm 1,\ \pm 2,\ \ldots,
$$
$$
= 2\pi \qquad \text{if } p = 0,
$$

the integral from $-\pi$ to π of $\exp[i(2k - m + n)\theta]$ vanishes unless $2k = m - n$. But in formula (5), $2k$ can be equal to $m - n$ only in case $m - n$ is zero or a positive even integer; that is, if $m - n = 2j$ $(j = 0, 1, 2, \ldots)$; this includes the condition that $m \geqq n$. In that case $k = j$ since $2k = m - n = 2j$; also, $m - k = 2j + n - j = n + j$ and, according to formula (5),

$$
(6) \qquad \int_{-\pi}^{\pi} c_{n+2j}(\theta)\, d\theta = 2\pi d_j d_{n+j} \qquad (j = 0, 1, 2, \ldots)
$$

but

$$
(7) \qquad \int_{-\pi}^{\pi} c_m(\theta)\, d\theta = 0 \qquad\qquad \text{if } m \neq n + 2j.
$$

Now the series in the representation

$$
\sum_{m=0}^{\infty} c_m(\theta) x^m = \alpha(x,\theta)\beta(x,\theta)
$$

is uniformly convergent with respect to θ over the interval $(-\pi,\pi)$. To show this, we note that, according to equation (5),

$$
|c_m(\theta) x^m| \leqq |x|^m \sum_{k=0}^{\infty} |d_k|\, |d_{m-k}|.
$$

The numbers represented by the right-hand member are independent of θ. They are the terms in the Cauchy product of two convergent series of positive terms, namely, the series of absolute values of the terms in the power series representations of $\alpha(x,0)$ and $\beta(x,0)$, series that are absolutely convergent. The Weierstrass test therefore applies to show the uniform convergence with respect to θ. Thus integration of the series is justified and, in view of formulas (6) and (7), we can write

$$\int_{-\pi}^{\pi} \alpha(x,\theta)\beta(x,\theta)\, d\theta = \sum_{m=0}^{\infty} x^m \int_{-\pi}^{\pi} c_m(\theta)\, d\theta$$

$$= 2\pi \sum_{j=0}^{\infty} d_j d_{n+j} x^{n+2j}.$$

Since $d_j = i^j/(j!\, 2^j)$, the last series becomes

$$i^n \sum_{j=0}^{\infty} \frac{(-1)^j}{j!(n+j)!} \left(\frac{x}{2}\right)^{n+2j} = i^n J_n(x).$$

Thus we have the integral representation

$$J_n(x) = \frac{i^{-n}}{2\pi} \int_{-\pi}^{\pi} \alpha(x,\theta)\beta(x,\theta)\, d\theta$$

which becomes, according to the definitions (3) of α and β,

$$(8) \quad J_n(x) = \frac{i^{-n}}{2\pi} \int_{-\pi}^{\pi} \exp(ix\cos\theta)e^{in\theta}\, d\theta \qquad (n = 0, 1, 2, \ldots).$$

Now $i = \exp(i\pi/2)$. Therefore

$$J_n(x) = \frac{1}{2\pi} \int_{-\pi}^{\pi} \exp(ix\cos\theta) \exp\left[in\left(\theta - \frac{\pi}{2}\right)\right] d\theta.$$

Then by substituting ϕ for $\frac{1}{2}\pi - \theta$ and noting that the integrand is periodic in θ and ϕ with period 2π, we find that

$$(9) \qquad J_n(x) = \frac{1}{2\pi} \int_{-\pi}^{\pi} \exp[i(x\sin\phi - n\phi)]\, d\phi.$$

The imaginary part of this integral vanishes, of course, since $J_n(x)$ is real when x is real. Therefore

$$J_n(x) = \frac{1}{2\pi} \int_{-\pi}^{\pi} \cos(x\sin\phi - n\phi)\, d\phi,$$

or since the integrand here is an even function of ϕ,

$$(10) \quad J_n(x) = \frac{1}{\pi} \int_{0}^{\pi} \cos(n\phi - x\sin\phi)\, d\phi \qquad (n = 0, 1, 2, \ldots).$$

This is *Bessel's integral form of J_n*.

73. Consequences of Bessel's Integral Form.   From the integral representation

$$(1) \quad J_n(x) = \frac{1}{\pi} \int_0^\pi \cos(n\phi - x\sin\phi)\, d\phi \qquad (n = 0, 1, 2, \ldots)$$

obtained above, it follows that

$$J_n'(x) = \frac{1}{\pi} \int_0^\pi \sin(n\phi - x\sin\phi)\sin\phi\, d\phi,$$

and so on, for $J_n''(x)$ and derivatives of higher order. Since the absolute value of the integrand of each of the integrals does not exceed unity, these *boundedness properties* can be seen from Bessel's integral form (1):

$$(2) \qquad\qquad |J_n(x)| \leqq 1, \qquad \left| \frac{d^k}{dx^k} J_n(x) \right| \leqq 1$$
$$(-\infty < x < \infty, k = 1, 2, \ldots).$$

The integrand in formula (1) can be written

$$\cos(n\phi - x\sin\phi) = f_n(x,\phi) + g_n(x,\phi),$$

where

$$f_n(x,\phi) = \cos n\phi \cos(x\sin\phi), \qquad g_n(x,\phi) = \sin n\phi \sin(x\sin\phi).$$

For each fixed x the graphs of f_n and g_n have these properties of symmetry with respect to the line $\phi = \frac{1}{2}\pi$:

$$(3) \qquad \begin{aligned} f_n(x, \pi - \phi) &= (-1)^n f_n(x,\phi), \\ g_n(x, \pi - \phi) &= -(-1)^n g_n(x,\phi). \end{aligned}$$

Consequently

$$\int_0^\pi f_{2n}\, d\phi = 2 \int_0^{\pi/2} f_{2n}\, d\phi, \qquad \int_0^\pi g_{2n}\, d\phi = 0,$$

and therefore

$$(4) \quad J_{2n}(x) = \frac{1}{\pi} \int_0^\pi \cos 2n\phi \cos(x\sin\phi)\, d\phi$$
$$= \frac{2}{\pi} \int_0^{\pi/2} \cos 2n\phi \cos(x\sin\phi)\, d\phi \quad (n = 0, 1, 2, \ldots).$$

Similarly, when $n = 1, 2, \ldots$, we see that

$$(5) \qquad J_{2n-1}(x) = \frac{1}{\pi} \int_0^\pi \sin\left[(2n-1)\phi\right] \sin\left(x \sin \phi\right) d\phi$$

$$= \frac{2}{\pi} \int_0^{\pi/2} \sin\left[(2n-1)\phi\right] \sin\left(x \sin \phi\right) d\phi.$$

As a special case of formula (4) we note that

$$(6) \qquad J_0(x) = \frac{2}{\pi} \int_0^{\pi/2} \cos\left(x \sin \phi\right) d\phi = \frac{2}{\pi} \int_0^{\pi/2} \cos\left(x \cos \theta\right) d\theta.$$

For each fixed x the Riemann-Lebesgue theorem (Lemma 1, Sec. 40) applies to the integrals in formulas (4) and (5) to show that $J_{2n}(x)$ and $J_{2n-1}(x)$ tend to zero as n increases. Thus

$$(7) \qquad\qquad\qquad \lim_{n \to \infty} J_n(x) = 0 \qquad (n = 0, 1, 2, \ldots).$$

A more important property is that, for each fixed n,

$$(8) \qquad\qquad\qquad \lim_{x \to \infty} J_n(x) = 0 \qquad (n = 0, 1, 2, \ldots).$$

To indicate a method of proving this, consider the special case $n = 0$. We substitute $t = \sin \phi$ in formula (6) to write

$$\frac{\pi}{2} J_0(x) = \int_0^c \frac{\cos xt}{(1 - t^2)^{\frac{1}{2}}} dt + \int_c^1 \frac{\cos xt}{(1 - t^2)^{\frac{1}{2}}} dt$$

where $0 < c < 1$. The second integral is improper but uniformly convergent with respect to x. Corresponding to any positive number ϵ, the absolute value of that integral can be made less than $\frac{1}{2}\epsilon$, uniformly for all x, by selecting c so that $1 - c$ is sufficiently small and positive. The Riemann-Lebesgue theorem then applies to the first integral, with that value of c. That is, there is a number x_ϵ such that the absolute value of the first integral is less than $\frac{1}{2}\epsilon$ whenever $x > x_\epsilon$. Therefore

$$\frac{\pi}{2} |J_0(x)| < \epsilon \qquad\qquad \text{when } x > x_\epsilon,$$

which proves property (8) when $n = 0$. The proof for the other cases is left to the problems.

PROBLEMS

1. Use integral forms of J_n to verify that (a) $J_0(0) = 1$; (b) $J_n(0) = 0$ when $n = 1, 2, \ldots$; (c) $J_0'(x) = -J_1(x)$.

2. Deduce from formula (4) that, when $n = 0, 1, 2, \ldots,$

$$J_{2n}(x) = (-1)^n \frac{2}{\pi} \int_0^{\pi/2} \cos(2n\theta) \cos(x \cos \theta) \, d\theta.$$

3. Deduce from formula (5) that, when $n = 1, 2, \ldots,$

$$J_{2n-1}(x) = -(-1)^n \frac{2}{\pi} \int_0^{\pi/2} \cos[(2n-1)\theta] \sin(x \cos \theta) \, d\theta.$$

4. Complete the proof of property (8) that

$$\lim_{x \to \infty} J_n(x) = 0 \qquad (n = 0, 1, 2, \ldots).$$

5. Verify directly from the representation

$$J_0(x) = \frac{1}{\pi} \int_0^\pi \cos(x \sin \phi) \, d\phi$$

that $J_0(x)$ satisfies Bessel's equation in which $n = 0$.

6. From integral representations of J_n prove that

$$\lim_{n \to \infty} n J_n(x) = 0 \qquad (n = 0, 1, 2, \ldots)$$

for each fixed x. (See Problem 4, Sec. 48.)

7. When $n = 1, 2, \ldots,$ show that

$$\int_0^\pi \sin(2n\phi) \sin(x \sin \phi) \, d\phi = 0.$$

Then with the aid of formula (5) and our Fourier theorem prove that, for all real x and θ,

$$\sin(x \sin \theta) = 2 \sum_{n=1}^\infty J_{2n-1}(x) \sin(2n-1)\theta.$$

8. Show that, when $n = 1, 2, \ldots,$

$$\int_0^\pi \cos[(2n-1)\phi] \cos(x \sin \phi) \, d\phi = 0.$$

Then with the aid of formula (4) and our Fourier theorem prove that for all real x and θ,

$$\cos(x \sin \theta) = J_0(x) + 2 \sum_{n=1}^\infty J_{2n}(x) \cos 2n\theta.$$

74. The Zeros of $J_0(x)$.   The integral form

$$\frac{\pi}{2} J_0(x) = \int_0^{\pi/2} \cos(x \sin \phi) \, d\phi$$

becomes, after the substitution $t = x \sin \phi$,

(1) $$\frac{\pi}{2} J_0(x) = \int_0^x \frac{\cos t}{\sqrt{x^2 - t^2}} \, dt \qquad (x > 0).$$

We shall show from the behavior of the integrand that this convergent improper integral has the value zero for an infinite sequence of positive values of x.

Let c_k denote odd multiples of $\pi/2$,

$$c_k = k\pi + \frac{\pi}{2} \qquad (k = 0, 1, 2, \ldots),$$

and consider the integral when $x = c_k$. Its integrand

(2) $$y_k(t) = \frac{\cos t}{(c_k^2 - t^2)^{\frac{1}{2}}} \qquad (0 \leq t < c_k)$$

has the limit zero as $t \to c_k$. We define $y_k(c_k)$ to be zero; then y_k is a continuous function of t over the interval of integration $0 \leq t \leq c_k$. Note that y_0 is defined only on the interval $0 \leq t \leq \pi/2$, y_1 on the interval $0 \leq t \leq 3\pi/2$, etc.

Now $y_k(t) > 0$ on the interval $0 < t < \pi/2$ $(k = 0, 1, 2, \ldots)$ since $\cos t > 0$ there. Similarly, we see that y_1, y_2, $\ldots$ have negative values over the interval $\pi/2 < t < 3\pi/2$, that y_2, y_3, $\ldots$ have positive values over the next interval of length π, and so on.

For a fixed positive integer k, let K_0 denote the area under the graph of $y_k(t)$ over the interval $(0, \frac{1}{2}\pi)$, which is the interval $0 < t < c_0$. Let the positive number K_j denote the area bounded by the t axis and that graph and the lines $t = c_{j-1}$ and $t = c_j$, where $j = 1, 2, \ldots, k$. Then

(3) $$\frac{\pi}{2} J_0(c_k) = \int_0^{c_k} y_k(t) \, dt = K_0 - K_1 + K_2$$
$$- \cdots + (-1)^k K_k.$$

Since c_k is fixed, the value of $(c_k^2 - t^2)^{\frac{1}{2}}$ diminishes when t increases toward c_k. But $|\cos t|$ is periodic with period π. It follows that $|y_k(t)|$ increases when t is increased by π and consequently that the areas K_j satisfy the inequalities $K_0 < K_1 < \cdots < K_k$. If k is an odd integer, then

$$\frac{\pi}{2} J_0(c_k) = -(K_1 - K_0) - (K_3 - K_2)$$
$$- \cdots - (K_k - K_{k-1}) < 0;$$

but if k is even,

$$\frac{\pi}{2} J_0(c_k) = K_0 + (K_2 - K_1) + \cdots + (K_k - K_{k-1}) > 0.$$

The continuous function $J_0(x)$ therefore takes on positive values at alternate points $x = \frac{1}{2}\pi$, $x = 2\pi + \frac{1}{2}\pi$, ... of the infinite set of points $x = k\pi + \frac{1}{2}\pi$ ($k = 0, 1, 2, \ldots$) and negative values at the other points of that set.   Consequently there is at least one point x_k on each interval between consecutive points of that set,

$$\left(k + \frac{1}{2}\right)\pi < x_k < \left(k + \frac{3}{2}\right)\pi \qquad (k = 0, 1, 2, \ldots),$$

such that $J_0(x_k) = 0$.

Actually, since J_0 is an analytic function of x which is not identically zero, it can have at most a finite number of zeros on any bounded interval.   Consequently *the positive roots of the equation $J_0(x) = 0$ consist of an infinite set of numbers x_m ($m = 1, 2, \ldots$) such that $x_m \to \infty$ as $m \to \infty$.*

Table 1 gives the values, correct to four significant figures, of the first five zeros of $J_0(x)$ and the corresponding values of $J_1(x)$. Extensive tables of numerical values of Bessel functions and their zeros will be found in books listed in the Bibliography.[1]

TABLE 1. $J_0(x_m) = 0$

m	1	2	3	4	5
x_m	2.405	5.520	8.654	11.79	14.93
$J_1(x_m)$	0.5191	−0.3403	0.2715	−0.2325	0.2065

75. Zeros of Other Functions. If for some pair of positive numbers b and c it is true that $J_n(b) = 0$ and $J_n(c) = 0$, then

---

[1] See the books by Jahnke *et al.*, Gray *et al.*, and Watson.

$x^{-n}J_n(x)$ also vanishes when $x = b$ and when $x = c$. It follows from Rolle's theorem that the derivative of $x^{-n}J_n(x)$ vanishes for at least one value of x between b and c. But (Sec. 71)

$$\frac{d}{dx}[x^{-n}J_n(x)] = -x^{-n}J_{n+1}(x) \qquad (n = 0, 1, 2, \ldots);$$

thus there is at least one zero of $J_{n+1}(x)$ between two positive zeros of $J_n(x)$. Again, J_{n+1} can have at most a finite number of zeros on each bounded interval because it is an analytic function for all x.

We have shown that the positive zeros of $J_0(x)$ constitute an unbounded infinite sequence of numbers. It now follows that the zeros of $J_1(x)$ must form such a set, then that the same is true

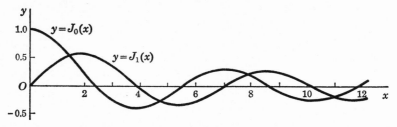

FIG. 22

for $J_2(x)$, etc. That is, for each fixed positive integer n the set of all positive roots of the equation $J_n(x) = 0$ is an infinite sequence $x = x_{nk}$ $(k = 1, 2, \ldots)$ where $x_{nk} \to \infty$ as $k \to \infty$.

Graphs of $J_0(x)$ and $J_1(x)$ are shown in Fig. 22.

The function $y = J_n(x)$ satisfies Bessel's equation, a linear homogeneous differential equation of the second order with the origin as a singular point. According to the uniqueness theorem (Sec. 19), there is just one solution that satisfies conditions $y(c) = y'(c) = 0$ where $c > 0$; that solution is identically zero. Consequently there is no positive number c such that $J_n(c) = J_n'(c) = 0$. That is, $J_n'(x)$ cannot vanish at a positive zero of $J_n(x)$; thus $J_n(x)$ must change its sign at that point.

Let b and c $(0 < b < c)$ be two consecutive zeros of $J_n(x)$. If $J_n'(b) > 0$, then $J_n(x) > 0$ when $b < x < c$ and $J_n(x)$ is decreasing at its zero c; that is, $J_n'(c) < 0$. Similarly, if $J_n'(b) < 0$, then $J_n'(c) > 0$ so that the values of J_n' alternate in sign at consecutive positive zeros of J_n.

We now consider the equation

(1) $$hJ_n(x) + xJ'_n(x) = 0,$$

where h is real and constant.   If b and c are consecutive positive zeros of $J_n(x)$, the function $hJ_n(x) + xJ'_n(x)$ has the values $bJ'_n(b)$ and $cJ'_n(c)$ at the points $x = b$ and $x = c$, respectively.   Since one of those values is positive and the other negative, that function vanishes at some point, or at some finite number of points, between b and c.   Equation (1) therefore has an infinite sequence of positive roots.   We collect our principal results as follows.

Theorem 2.   *For each fixed* n $(n = 0, 1, 2, \ldots)$ *the set of all positive roots of the equation*

(2) $$J_n(x) = 0$$

consists of an infinite sequence $x = x_j$ $(j = 1, 2, \ldots)$ *such that* $x_j \to \infty$ *as* $j \to \infty$; *also the set of all positive roots of equation* (1), *where* h *is a constant, including zero, is always a sequence of that type.*

We note that $x = 0$ is a root of both equations (1) and (2) if n is a positive integer.   It is also a root of the equation $J'_0(x) = 0$.

When $x = b$ is a root of equation (2), then $x = -b$ is also a root since $J_n(-b) = (-1)^n J_n(b)$.   That statement is true of equation (1) as well, for with the aid of the formula (Sec. 71)

$$xJ'_n(x) = nJ_n(x) - xJ_{n+1}(x)$$

equation (1) can be written

(3) $$(h + n)J_n(x) - xJ_{n+1}(x) = 0.$$

76. Orthogonal Sets of Bessel Functions.   Bessel's equation

$$t^2 \frac{d^2y}{dt^2} + t\frac{dy}{dt} + (t^2 - n^2)y = 0 \qquad (n = 0, 1, 2, \ldots)$$

has the particular solution $y = J_n(t)$.   As pointed out in Sec. 68, using different notation, in terms of the variables x and X, where $t = \lambda x$ and $y(t) = X(x)$, the equation takes the form

$$x^2 X''(x) + xX'(x) + (\lambda^2 x^2 - n^2)X(x) = 0.$$

The self-adjoint form of this equation is

$$(1) \qquad (xX')' + \left(\lambda^2 x - \frac{n^2}{x}\right) X = 0,$$

and $X(x) = J_n(\lambda x)$ is a particular solution.

For each fixed n $(n = 0, 1, 2, \ldots)$ equation (1) is a special case of the Sturm-Liouville equation (4), Sec. 32, with the eigenvalue parameter written as λ^2 instead of λ.   In this special case $r(x) = p(x) = x$ and $q(x) = -n^2/x$.

Consider the singular Sturm-Liouville problem, on an interval $(0,c)$, consisting of equation (1), a boundary condition

$$(2) \qquad b_1 X(c) + b_2 X'(c) = 0,$$

and the requirement that X and X' shall be continuous over the closed interval $0 \leq x \leq c$.   The constants b_1 and b_2 are real. The end point $x = 0$ is the singular point of the differential equation; also, q is discontinuous there unless $n = 0$.

This problem has an infinite sequence of real eigenvalues for λ, as we can now show.   The solution of equation (1) which satisfies the above continuity requirements is, except for an arbitrary constant factor, $X = J_n(\lambda x)$.

Let us agree that *the symbol $J_n'(\lambda x)$ stands for the derivative of J_n with respect to the argument of J_n*:

$$J_n'(\lambda x) = \frac{d}{d(\lambda x)} J_n(\lambda x).$$

Then $d/dx \, [J_n(\lambda x)] = \lambda J_n'(\lambda x)$ and condition (2) becomes

$$(3) \qquad b_1 J_n(\lambda c) + b_2 \lambda J_n'(\lambda c) = 0.$$

In the important special case $b_2 = 0$ the condition is

$$(4) \qquad J_n(\lambda c) = 0.$$

When $b_2 \neq 0$, we may multiply through by c/b_2 to write condition (3) in the form

$$(5) \qquad h J_n(\lambda c) + \lambda c J_n'(\lambda c) = 0.$$

We have shown in the preceding section that there are infinite sequences of positive values of λc that satisfy equations (4) and (5).   If we designate each sequence by x_j $(j = 1, 2, \ldots)$, but keep in mind that *the numbers x_j depend on the values of n, b_1, and b_2,* then λ is an eigenvalue if $\lambda c = x_j$; that is,

$$(6) \qquad\qquad \lambda_j = \frac{x_j}{c} \qquad\qquad (j = 1, 2, \ldots)$$

are real eigenvalues. The corresponding eigenfunctions are

$$(7) \qquad\qquad X_j(x) = J_n(\lambda_j x) \qquad\qquad (j = 1, 2, \ldots).$$

Each negative root $\lambda = -\lambda_j$ of equations (4) or (5) gives the same eigenfunction, except for the factor $(-1)^n$, as the corresponding positive root λ_j.

When $n = 1, 2, \ldots$, zero is not an eigenvalue because $J_n(\lambda x)$ is identically zero if $\lambda = 0$. If $n = 0$, then $\lambda = 0$ is not a solution of equation (4), nor is it a solution of equation (5) unless $h = 0$ also.

When $n = h = 0$, equation (5) can be written $J_0'(\lambda c) = 0$ or $J_1(\lambda c) = 0$, and the eigenfunction corresponding to the root $\lambda = 0$ is $J_0(0)$; that is

$$X(x) = 1 \qquad \text{when } \lambda = 0 \qquad (n = h = 0).$$

This is the only case in which $\lambda = 0$ is an eigenvalue.

According to the power series representation of J_n, the function $i^{-n}J_n(ix)$ has only positive real values when $x > 0$ (Problem 4, Sec. 71). It follows that equation (4) has no purely imaginary roots $\lambda = i\mu$ (μ real).

From now on we assume that $h \geqq 0$ in equation (5), as is usually the case in the applications. Since that equation can be written in the form (3), Sec. 75,

$$(8) \qquad\qquad (h + n)J_n(\lambda c) - \lambda c J_{n+1}(\lambda c) = 0 \qquad\qquad (h \geqq 0)$$

and since $h + n \geqq 0$, the power series representing J_n and J_{n+1} show that the equation has no purely imaginary roots λ.

Thus the eigenvalue problem consisting of equations (1) and (2), where $b_1 b_2 \geqq 0$, has no purely imaginary eigenvalues λ.

Cases (a) of Theorems 3 and 4, Sec. 33, give conditions under which our singular Sturm-Liouville problem has orthogonal eigenfunctions and real eigenvalues λ^2. We note that the proofs of the theorems do not depend on the continuity of $q(x)$ at the end point $x = a$ ($a = 0$ in our special case) as long as X and X' are continuous over the closed interval. That continuity requires that the solution $X = J_n(\lambda x)$ of equation (1) be used even if λ is complex since $Y_n(\lambda x)$ has the same type of discontinuity as $\log \lambda x$ at $x = 0$, while the convergence of the power series for $J_n(\lambda x)$ and $J_n'(\lambda x)$ ensures the continuity of those functions for all values of λx.

According to Theorem 4, Sec. 33, all eigenvalues of λ^2 are real. We have seen that λ itself cannot be purely imaginary; hence the eigenvalues of λ are all real. And those real values are all represented by the numbers λ_j given by equation (6), where $\lambda_j > 0$, except that $\lambda_1 = 0$ in one special case, $n = h = 0$. Let us agree to arrange those eigenvalues in ascending order of magnitude so that $\lambda_j < \lambda_{j+1}$.

Theorem 3, Sec. 33, gives the orthogonality property

$$(9) \qquad \int_0^c x J_n(\lambda_j x) J_n(\lambda_k x) \, dx = 0 \qquad\qquad (\lambda_j \neq \lambda_k).$$

Note that this orthogonality of the eigenfunctions with weight function x, on the interval $(0,c)$, is the same as ordinary orthogonality of the functions $\sqrt{x} J_n(\lambda_j x)$. Also note that many orthogonal sets are represented here, depending on the values of n, b_1, b_2, and c.

We summarize our results in the following theorem.

Theorem 3. *Let n have one of the values $0, 1, 2, \ldots$. The set of functions $\{J_n(\lambda_j x)\}$ $(j = 1, 2, \ldots)$ is orthogonal on the interval $(0,c)$ with weight function x when $\lambda = \lambda_j$ are either (a) the positive roots of the equation $J_n(\lambda c) = 0$, or (b) the positive roots of equation (5) where $h \geqq 0$ and h and n are not both zero, or (c) in case $h = n = 0$, then $\lambda_1 = 0$ and $\lambda_2, \lambda_3, \ldots$ are the positive roots of the equation $J_0'(\lambda c) = 0$.*

Those numbers λ_j are all the eigenvalues of λ, and $X = J_n(\lambda_j x)$ are the eigenfunctions, of the singular Sturm-Liouville problem consisting of Bessel's equation (1) together with the boundary condition

$$(10) \qquad\qquad X(c) = 0$$

in case (a); or in case (b), the boundary condition

$$(11) \quad hX(c) + cX'(c) = 0 \qquad\qquad (h \geqq 0, h^2 + n^2 > 0);$$

or in case (c), where $n = 0$, the condition

$$(12) \qquad\qquad X'(c) = 0.$$

77. The Orthonormal Functions.   The function

$$X(x) = J_n(\lambda x)$$

satisfies the equation

(1) $$(xX')' + \left(\lambda^2 x - \frac{n^2}{x}\right) X = 0.$$

We multiply the terms by the factor $2xX'$ to write

$$\frac{d}{dx}(xX')^2 + (\lambda^2 x^2 - n^2)\frac{d}{dx}(X^2) = 0.$$

After integrating both terms here and using integration by parts in the second term we find that

$$[(xX')^2 + (\lambda^2 x^2 - n^2)X^2]_0^c = 2\lambda^2 \int_0^c xX^2\,dx.$$

Since $X = J_n(\lambda x)$, where $n = 0, 1, 2, \ldots$, the quantity inside the brackets vanishes when $x = 0$, and the equation can be written

(2) $$2\lambda^2 \int_0^c x[J_n(\lambda x)]^2\,dx = \lambda^2 c^2[J_n'(\lambda c)]^2 + (\lambda^2 c^2 - n^2)[J_n(\lambda c)]^2.$$

Except in the one case where $\lambda = 0$, formula (2) gives the norms of our eigenfunctions.

When λ_j are the positive roots of the equation

(3) $$J_n(\lambda c) = 0,$$

formula (2) becomes

$$2 \int_0^c x[J_n(\lambda_j x)]^2\,dx = c^2[J_n'(\lambda_j c)]^2.$$

But we have seen that

$$rJ_n'(r) = nJ_n(r) - rJ_{n+1}(r)$$

and therefore $\lambda_j c J_n'(\lambda_j c) = -\lambda_j c J_{n+1}(\lambda_j c)$, hence

(4) $$\|J_n(\lambda_j x)\|^2 = \frac{c^2}{2}[J_{n+1}(\lambda_j c)]^2 \quad (j = 1, 2, \ldots).$$

When λ_j are the positive roots of the equation

$$(5) \qquad\qquad hJ_n(\lambda c) + \lambda cJ_n'(\lambda c) = 0 \qquad\qquad (h \geq 0),$$

we find from equation (2) that

$$(6) \qquad\qquad \|J_n(\lambda_j x)\|^2 = \frac{\lambda_j^2 c^2 - n^2 + h^2}{2\lambda_j^2}\, [J_n(\lambda_j c)]^2.$$

But in case $n = h = 0$, then $\lambda_1 = 0$ and $J_0(\lambda_1 x) = 1$; thus

$$(7) \qquad\qquad \|J_0(\lambda_1 x)\|^2 = \int_0^c x\, dx = \tfrac{1}{2}c^2.$$

The normalized eigenfunctions of our singular Sturm-Liouville problem can now be written

$$(8) \qquad\qquad \phi_{nj}(x) = \frac{J_n(\lambda_j x)}{\|J_n(\lambda_j x)\|} \qquad\qquad (j = 1, 2, \ldots).$$

The norms here are given for the three types of boundary conditions by formulas (4), (6), and (7); and the set of eigenvalues λ_j depends on the particular boundary condition used and on the value of n. The set (8) is orthonormal on the interval $(0,c)$ with weight function x:

$$\int_0^c x\phi_{nj}(x)\phi_{nk}(x)\, dx = \begin{cases} 0 & \text{if } j \neq k \\ 1 & \text{if } j = k. \end{cases}$$

78. Fourier-Bessel Series. Since the functions $\phi_{nj}(x)$ constitute all the normalized eigenfunctions of our singular Sturm-Liouville problem, we can anticipate a representation of functions f on the interval $(0,c)$ by the generalized Fourier series of those orthonormal functions.

Let c_{nj} $(j = 1, 2, \ldots)$ denote the Fourier constants of a function f with respect to ϕ_{nj} on the interval. Then

$$c_{nj} = \int_0^c x\phi_{nj}(x)f(x)\, dx = \frac{1}{\|J_n(\lambda_j x)\|}\int_0^c xJ_n(\lambda_j x)f(x)\, dx,$$

and the generalized Fourier series corresponding to f is

$$(1) \qquad \sum_{j=1}^{\infty} c_{nj}\phi_{nj}(x) = \sum_{j=1}^{\infty} \frac{J_n(\lambda_j x)}{\|J_n(\lambda_j x)\|^2}\int_0^c \xi J_n(\lambda_j \xi)f(\xi)\, d\xi.$$

In view of the formulas for the norms, found in the preceding section, series (1) can be written

$$(2) \qquad \sum_{j=1}^{\infty} A_{nj} J_n(\lambda_j x) \sim f(x) \qquad (0 < x < c),$$

where the coefficients A_{nj} have the following values.

When λ_j are the positive roots of the equation

$$(3) \qquad J_n(\lambda c) = 0 \qquad (n = 0, 1, 2, \ldots),$$

$$(4) \qquad A_{nj} = \frac{2}{c^2 [J_{n+1}(\lambda_j c)]^2} \int_0^c x J_n(\lambda_j x) f(x)\, dx.$$

When λ_j are the positive roots of the equation

$$(5) \quad h J_n(\lambda c) + \lambda c J_n'(\lambda c) = 0 \qquad (h \geqq 0, n = 0, 1, 2, \ldots),$$

$$(6) \quad A_{nj} = \frac{2\lambda_j^2}{(\lambda_j^2 c^2 - n^2 + h^2)[J_n(\lambda_j c)]^2} \int_0^c x J_n(\lambda_j x) f(x)\, dx.$$

In the special case $n = h = 0$, however, the root $\lambda = 0$ of equation (5) is also an eigenvalue. We then write $\lambda_1 = 0$, $J_0(\lambda_1 x) = 1$ and the first term in series (2) is this constant:

$$(7) \qquad A_{01} = \frac{2}{c^2} \int_0^c x f(x)\, dx.$$

Proofs that series (2) does converge to $f(x)$, under conditions on f that ensure the representation of that function by its Fourier sine or cosine series on the interval $(0, c)$, usually involve the theory of functions of a complex variable. We state, without proof, one form of such a representation theorem.[1]

Theorem 4. *If f is a function which, together with its derivative f', is sectionally continuous on the interval $(0, c)$, then series (2) converges to the mean value of f at each interior point of the interval. That is,*

$$(8) \quad \tfrac{1}{2}[f(x + 0) + f(x - 0)] = \sum_{j=1}^{\infty} A_{nj} J_n(\lambda_j x) \qquad (0 < x < c),$$

[1] Proved in Watson's "Theory of Bessel Functions"; also, see p. 73 of Titchmarsh, "Eigenfunction Expansions," 1946. Watson's book gives the most extensive treatment of Bessel functions of various kinds; also, see the books by Gray *et al.* and Bowman.

where the coefficients A_{nj} are defined by equation (4) *or* (6) *or* (7), *depending on the particular equation that determines the roots* λ_j.

The expansion (8) is known as the *Fourier-Bessel series* representation of $f(x)$.

The results stated in Theorems 2, 3, and 4 are also valid when n is replaced by an arbitrary positive number ν although we have not developed the properties of the functions J_ν far enough to establish this fact.

For functions on the unbounded interval $x > 0$ there is an integral representation in terms of J_ν corresponding to the Fourier sine or cosine integral formula.[1] The representation, for a fixed ν $(\nu \geqq -\frac{1}{2})$,

$$f(x) = \int_0^\infty \alpha J_\nu(\alpha x) \int_0^\infty \xi J_\nu(\alpha \xi) f(\xi) \, d\xi \, d\alpha \qquad (x > 0),$$

known as *Hankel's integral formula*, is valid in case f and f' are sectionally continuous on each bounded interval and if $\sqrt{x} f(x)$ is absolutely integrable from zero to infinity and $f(x)$ is defined as its mean value at each point of discontinuity.

If the interval $(0,c)$ is replaced by some interval (a,b), where $0 < a < b$, the Sturm-Liouville problem consisting of Bessel's equation (1), Sec. 76, and a linear homogeneous condition of type (2), Sec. 76, at each end point of the interval, is no longer singular. The problem is simply a special case of the Sturm-Liouville problem. The eigenfunctions are linear combinations of the Bessel functions J_n and Y_n.

PROBLEMS

1. For the singular Sturm-Liouville problem

$$(xX')' + \lambda^2 xX = 0 \qquad (0 < x < 2),$$
$$X(2) = 0,$$

show that the eigenvalues $\lambda = \lambda_j$ are the positive roots of the equation $J_0(2\lambda) = 0$ and that the eigenfunctions are $X_j = J_0(\lambda_j x)$. Also, give approximate numerical values of the first few eigenvalues.

$Ans.$ $\lambda_1 = 1.2, \lambda_2 = 2.8, \ldots$

---

[1] Sneddon, "Fourier Transforms," chap. 2.   See chap. 7, vol. 2, "Higher Transcendental Functions," by Erdélyi *et al.* for a summary of representations in terms of Bessel functions.

2. For the eigenvalue problem

$$(xX')' + \lambda^2 xX = 0 \qquad (0 < x < c),$$
$$X'(c) = 0,$$

show that $\lambda_1 = 0$, that $\lambda_2, \lambda_3, \ldots$ are the positive roots of the equation $J_1(\lambda c) = 0$, and that $X_1(x) = 1$ and $X_j(x) = J_0(\lambda_j x)$ when $j = 2$, $3, \ldots$.

3. Represent the function $f(x) = 1 \quad (0 < x < c)$ in series of the functions $J_0(\lambda_j x)$, where λ_j are the positive roots of the equation $J_0(\lambda c) = 0$.

$$Ans. \quad 1 = \frac{2}{c} \sum_{j=1}^{\infty} \frac{J_0(\lambda_j x)}{\lambda_j J_1(\lambda_j c)} \quad (0 < x < c).$$

4. In the representation of the function $f(x) = 1 \quad (0 < x < c)$ in series of the functions $J_0(\lambda_j x)$, where λ_j are the nonnegative roots of $J_0'(\lambda c) = 0$, show that $A_1 = 1$ and $A_j = 0$ when $j = 2, 3, \ldots$.

5. If $f(x) = 1$ when $0 < x < 1$ and $f(x) = 0$ when $1 < x < 2$, and $f(1) = \frac{1}{2}$, and if λ_j are the positive roots of the equation $J_0(2\lambda) = 0$, obtain the representation

$$f(x) = \frac{1}{2} \sum_{j=1}^{\infty} \frac{J_1(\lambda_j)}{\lambda_j [J_1(2\lambda_j)]^2} J_0(\lambda_j x) \qquad (0 < x < 2).$$

6. For the eigenvalue problem

$$(xX')' + \left(\lambda^2 x - \frac{1}{x}\right) X = 0 \qquad (0 < x < 1),$$
$$X(1) = 0,$$

show that λ_j are the positive roots of the equation $J_1(\lambda) = 0$ and that $X_j = J_1(\lambda_j x)$. Then obtain this representation of the function $f(x) = x \quad (0 \le x < 1)$ in series of those eigenfunctions:

$$x = 2 \sum_{j=0}^{\infty} \frac{J_1(\lambda_j x)}{\lambda_j J_2(\lambda_j)} \quad (0 \le x < 1).$$

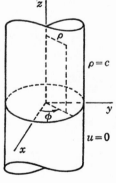

79. Temperatures in a Long Cylinder. Let the lateral surface $\rho = c$ of an infinitely long circular cylinder (Fig. 23), or a cylinder of finite altitude with insulated bases, be kept at temperature zero. The initial temperature distribution is a function

FIG. 23

$f(\rho)$ of distance from the axis, only. We shall derive a formula for the temperatures $u(\rho,t)$ in the cylinder.

The heat equation in cylindrical coordinates, and the boundary conditions, are

$$(1) \qquad \frac{\partial u}{\partial t} = k\left(\frac{\partial^2 u}{\partial \rho^2} + \frac{1}{\rho}\frac{\partial u}{\partial \rho}\right) \quad (0 < \rho < c, t > 0),$$

$$(2) \qquad u(c - 0, t) = 0 \qquad\qquad (t > 0),$$

$$(3) \qquad u(\rho, + 0) = f(\rho) \qquad\qquad (0 < \rho < c).$$

Also, when $t > 0$, the function u is to be continuous throughout the cylinder, in particular at the axis $\rho = 0$. We assume that f and f' are sectionally continuous on the interval $(0,c)$ and, for convenience, that f is defined as the mean value of its limits at each point of discontinuity.

Particular solutions of the homogeneous conditions (1) and (2), of the type $R(\rho)T(t)$, must satisfy the equations

$$RT' = kT(R'' + \rho^{-1} R), \qquad R(c)T(t) = 0.$$

Separating variables in the first equation, we have

$$\frac{T'}{kT} = \frac{1}{R}\left(R'' + \frac{R'}{\rho}\right) = -\lambda^2,$$

where $-\lambda^2$ is some constant yet to be specified. Thus

$$(4) \qquad T'(t) + \lambda^2 kT(t) = 0$$

and

$$\rho R''(\rho) + R'(\rho) + \lambda^2 \rho R(\rho) = 0.$$

The function $R(\rho)$ must therefore satisfy the conditions

$$(5) \qquad (\rho R')' + \lambda^2 \rho R = 0 \qquad (0 < \rho < c), \qquad R(c) = 0.$$

The differential equation in R is Bessel's equation with the parameter λ, in which $n = 0$. Problem (5), together with continuity conditions on R over the interval $0 \le \rho \le c$, is a special case of the singular Sturm-Liouville problem consisting of equations (1) and (2) in Sec. 76. According to case (a) of Theorem 3, the eigenvalues $\lambda = \lambda_j$ of problem (5) are the positive roots of the equation

$$(6) \qquad J_0(\lambda c) = 0$$

and $R = J_0(\lambda_j\rho)$ are the eigenfunctions.

When $\lambda = \lambda_j$, we see from equation (4) that $T = \exp(-\lambda_j{}^2 kt)$ so that, except for a constant factor,

$$R(\rho)T(t) = J_0(\lambda_j\rho) \exp(-\lambda_j{}^2 kt) \qquad (j = 1, 2, \ldots)$$

The extended linear combination of those functions

$$(7) \qquad u(\rho,t) = \sum_{j=1}^{\infty} A_j J_0(\lambda_j\rho) \exp(-\lambda_j{}^2 kt)$$

formally satisfies the homogeneous conditions (1) and (2) in our boundary value problem. It also satisfies the nonhomogeneous initial condition (3) if $u(\rho, +0) = u(\rho,0)$ and if the coefficients A_j can be determined so that

$$f(\rho) = \sum_{j=1}^{\infty} A_j J_0(\lambda_j\rho) \qquad (0 < \rho < c).$$

This representation is valid, according to the Fourier-Bessel series (Theorem 4), if the coefficients have the values

$$(8) \quad A_j = \frac{2}{c^2[J_1(\lambda_j c)]^2} \int_0^c \xi J_0(\lambda_j\xi)f(\xi)\, d\xi \qquad (j = 1, 2, \ldots).$$

The formal solution of the boundary value problem is therefore represented by formula (7) with the coefficients (8), where λ_j are all positive roots of equation (6). That is, our temperature formula can be written as

$$u(\rho,t) = \frac{2}{c^2} \sum_{j=1}^{\infty} \frac{J_0(\lambda_j\rho)}{[J_1(\lambda_j c)]^2} \exp(-\lambda_j{}^2 kt) \int_0^c \xi J_0(\lambda_j\xi)f(\xi)\, d\xi.$$

Verification. We can establish formula (7) as a solution of our problem by the procedure followed in Sec. 61 if we use two additional properties of zeros of Bessel functions, namely that $\lambda_{j+1} - \lambda_j \to \pi/c$ as $j \to \infty$, and the numbers $[\sqrt{\lambda_j}J_1(\lambda_j c)]^{-1}$ are bounded for all j $(j = 1, 2, \ldots)$.[1]

[1] Those properties follow from an asymptotic representation of $J_n(x)$ for large values of x, which can be written

$$\sqrt{\pi x}\, J_n(x) = \sqrt{2} \cos(x - \tfrac{1}{2}n\pi - \tfrac{1}{4}\pi) + x^{-1}\theta_n(x),$$

where $\theta_n(x)$ is bounded as x tends to infinity. For derivations, see Courant and Hilbert's, "Methods of Mathematical Physics," vol. 1, p. 526, or Watson, "Theory of Bessel Functions."

Since f and J_0 are bounded, it follows from formula (8) that the numbers A_j/λ_j are bounded. Therefore a number M exists such that, for each positive number t_0, the absolute values of the terms of series (7) are less than the constant terms

$$M\lambda_j \exp(-\lambda_j^2 k t_0)$$

when $0 \leqq \rho \leqq c$ and $t \geqq t_0$. The series of those constant terms converges because $\lambda_{j+1} - \lambda_j \to \pi/c$ as $j \to \infty$. Series (7) therefore converges uniformly with respect to ρ and t $(t \geqq t_0)$, and its sum $u(\rho,t)$ is a continuous function of its two variables when $0 \leqq \rho \leqq c$ and $t > 0$. But $u(c,t) = 0$; hence condition (2) is satisfied.

The derivatives of $J_0(x)$ are also bounded, and the series of constants $M\lambda_j^m \exp(-\lambda_j^2 k t_0)$, where $m = 2, 3$, converges. So it follows in the same way that the differentiated series converge uniformly when $t \geqq t_0$, and hence that the function (7) satisfies the heat equation (1).

Finally, owing to the convergence of series (7) to $f(\rho)$ when $t = 0$, Abel's test applies to show that $u(\rho, +0) = u(\rho,0)$, when $0 < \rho < c$. Condition (3) is therefore satisfied, and our solution is established.

80. Heat Transfer at the Surface of the Cylinder. Let us replace the condition that the surface of the infinite cylinder is at temperature zero by a condition that heat transfer takes place there into surroundings at temperature zero. We assume a linear law of surface heat transfer, sometimes called Newton's law: the flux through the surface is proportional to the difference of temperatures of the surface and its surroundings. Thus if K is the conductivity of the material in the cylinder and E the external conductivity or emissivity at the surface $\rho = c$,

$$-Ku_\rho(c,t) = Eu(c,t).$$

We assume that K and E are constants $(K > 0, E \geqq 0)$, and write $h = cE/K$.

The boundary value problem for the temperature function u is

(1) $\qquad u_t(\rho,t) = k[u_{\rho\rho}(\rho,t) + \rho^{-1}u_\rho(\rho,t)] \qquad (0 < \rho < c,\ t > 0),$

(2) $\qquad\qquad\qquad cu_\rho(c,t) = -hu(c,t) \qquad\qquad\qquad (t > 0),$

(3) $\qquad\qquad\qquad u(\rho,0) = f(\rho) \qquad\qquad\qquad\qquad (0 < \rho < c).$

Separation of variables now produces this eigenvalue problem in the function $R(\rho)$:

(4)
$$(\rho R')' + \lambda^2 \rho R = 0 \qquad (0 < \rho < c),$$
$$hR(c) + cR'(c) = 0.$$

Thus $R = J_0(\lambda \rho)$, where λ must satisfy the equation

(5)
$$hJ_0(\lambda c) + \lambda c J_0'(\lambda c) = 0.$$

In case $h > 0$, then λ_j are the positive roots of the equation (5), and solutions of the homogeneous conditions are

$$R(\rho)T(t) = J_0(\lambda_j \rho) \exp (-\lambda_j^2 kt) \qquad (j = 1, 2, \ldots).$$

The formal solution of our problem is then

(6)
$$u(\rho,t) = \sum_{j=1}^{\infty} A_j J_0(\lambda_j \rho) \exp (-\lambda_j^2 kt),$$

where, according to the initial condition (3) and Theorem 4,

(7)
$$A_j = \frac{2\lambda_j^2}{(\lambda_j^2 c^2 + h^2)[J_0(\lambda_j c)]^2} \int_0^c \xi J_0(\lambda_j \xi)f(\xi)\, d\xi$$
$$(j = 1, 2, \ldots).$$

In case $h = 0$, then $\lambda_1 = 0$ and $J_0(\lambda_1 \rho) = 1$, and $\lambda_2, \lambda_3, \ldots$ are the positive roots of the equation $J_1(\lambda c) = 0$. The temperature formula becomes

(8)
$$u(\rho,t) = A_1 + \sum_{j=2}^{\infty} A_j J_0(\lambda_j \rho) \exp (-\lambda_j^2 kt),$$

where (Theorem 4)

(9)
$$A_1 = \frac{2}{c^2} \int_0^c \xi f(\xi)\, d\xi,$$

(10)
$$A_j = \frac{2}{c^2[J_0(\lambda_j c)]^2} \int_0^c \xi J_0(\lambda_j \xi)f(\xi)\, d\xi \qquad (j = 2, 3, \ldots).$$

This is the case in which the surface $\rho = c$ is thermally insulated.

PROBLEMS

1. Write the formula for the temperatures $u(\rho,t)$ in an infinite cylinder $\rho \leqq 1$ under the conditions $u(1,t) = 0$, $u(\rho,0) = u_0$, where u_0 is a con-

stant.   Give approximate numerical values of the first few coefficients
in the series.

Ans. $u = 2u_0[0.80J_0(2.4\rho) \exp(-5.8kt) - 0.53J_0(5.5\rho) \exp(-30kt)$
$$+ 0.43J_0(8.6\rho) \exp(-75kt) - \cdots].$$

2. Over a long solid cylinder $\rho \leqq 1$, at uniform temperature A, is
tightly fitted a long hollow cylinder $1 \leqq \rho \leqq 2$ of the same material at
temperature B.   The outer surface $\rho = 2$ is then kept at temperature
B.   Derive this formula for the temperatures in the cylinder of radius 2
so formed:

$$u(\rho,t) = B + \frac{A - B}{2} \sum_{j=1}^{\infty} \frac{J_1(\lambda_j)}{\lambda_j[J_1(2\lambda_j)]^2} J_0(\lambda_j\rho) \exp(-\lambda_j^2 kt),$$

where λ_j are the positive roots of $J_0(2\lambda) = 0$.   This is a temperature
problem in shrunken fittings.

3. A function $V(\rho,z)$ is harmonic interior to the cylinder bounded by
the surfaces $\rho = c$, $z = 0$, and $z = b$.   If $V = 0$ on the first two surfaces
and $V(\rho,b) = f(\rho)$ $(0 < \rho < c)$, derive the formula

$$V(\rho,z) = \sum_{j=1}^{\infty} A_j J_0(\lambda_j\rho) \frac{\sinh \lambda_j z}{\sinh \lambda_j b},$$

where λ_j are the positive roots of $J_0(\lambda c) = 0$ and the coefficients A_j are
given by formula (8), Sec. 79.

4. Derive a formula for the steady temperatures $u(\rho,z)$ in the solid
cylinder bounded by the surfaces $\rho = 1$, $z = 0$, and $z = 1$ if $u = 0$ on
the surface $\rho = 1$, $u = 1$ on the base $z = 1$, and the base $z = 0$ is
insulated.

$$Ans.\quad u = 2 \sum_{j=1}^{\infty} \frac{J_0(\lambda_j\rho)}{\lambda_j J_1(\lambda_j)} \frac{\cosh \lambda_j z}{\cosh \lambda_j}, \qquad [J_0(\lambda_j) = 0, \lambda_j > 0].$$

5. A solid cylinder is bounded by the surfaces $\rho = 1$, $z = 0$, and $z = b$.
The first surface is insulated, the second kept at temperature zero, and
the last at temperatures $f(\rho)$.   Derive this formula for the steady temperatures $u(\rho,z)$ in the cylinder

$$u = \frac{2z}{b} \int_0^1 \xi f(\xi)\, d\xi + 2 \sum_{j=2}^{\infty} \frac{J_0(\lambda_j\rho)}{[J_0(\lambda_j)]^2} \frac{\sinh \lambda_j z}{\sinh \lambda_j b} \int_0^1 \xi J_0(\lambda_j\xi) f(\xi)\, d\xi,$$

where $\lambda_2, \lambda_3, \ldots$ are the positive roots of $J_1(\lambda) = 0$.

6. When $f(\rho) = A$ $(0 < \rho < 1)$ in Problem 5, show that $u(\rho,z) = Az/b$.

7. Find the bounded steady temperatures $u(\rho,z)$ in the semi-infinite
cylinder $\rho \leqq 1$, $z \geqq 0$ if $u = 1$ on the base $z = 0$, while heat transfer

into surroundings at temperature zero takes place, according to the
linear law (Sec. 80), at the surface $\rho = 1$.

$$Ans. \quad u = 2h \sum_{j=1}^{\infty} \frac{J_0(\lambda_j\rho)}{J_0(\lambda_j)} \frac{\exp{(-\lambda_j z)}}{\lambda_j{}^2 + h^2}, \quad [hJ_0(\lambda_j) = \lambda_j J_1(\lambda_j)].$$

8. Solve this boundary value problem for $u(x,t)$:

$$\begin{aligned}
xu_t &= (xu_x)_x - n^2 x^{-1}u & (0 < x < c,\, t > 0), \\
u(c,t) &= 0 & (t > 0), \\
u(x,0) &= f(x) & (0 < x < c),
\end{aligned}$$

where u is continuous when $0 \leqq x \leqq c$ and $t > 0$ and the constant n is a
nonnegative integer.

$$Ans. \quad u = \sum_{j=1}^{\infty} A_{nj} J_n(\lambda_j x) \exp{(-\lambda_j{}^2 t)}, \text{ where } \lambda_j \text{ and } A_{nj} \text{ have the}$$

values indicated by formulas (3) and (4), Sec. 78.

81. Vibration of a Circular Membrane. A membrane,
stretched over a fixed circular frame $\rho = c$ in the plane $z = 0$,
is given an initial displacement $z = f(\rho,\phi)$ and released from rest
in that position. Its transverse displacements $z(\rho,\phi,t)$, where
ρ, ϕ, and z are cylindrical coordinates, will be found as the continuous function that satisfies this boundary value problem:

$$(1) \qquad z_{tt} = a^2(z_{\rho\rho} + \rho^{-1}z_\rho + \rho^{-2}z_{\phi\phi}),$$
$$(2) \qquad z(c,\phi,t) = 0 \qquad (-\pi \leqq \phi \leqq \pi,\, t \geqq 0),$$
$$(3) \qquad z_t(\rho,\phi,0) = 0, \qquad z(\rho,\phi,0) = f(\rho,\phi)$$
$$(0 \leqq \rho \leqq c,\, -\pi \leqq \phi \leqq \pi).$$

A function $z = R(\rho)\Phi(\phi)T(t)$ satisfies equation (1) if

$$(4) \qquad \frac{T''}{a^2 T} = \frac{1}{R}\left(R'' + \frac{R'}{\rho}\right) + \frac{1}{\rho^2}\frac{\Phi''}{\Phi} = -\lambda^2,$$

where $-\lambda^2$ is any constant. We separate variables again in
equations (4) and write $\Phi''/\Phi = -\mu$. Then we find that the
function $R\Phi T$ satisfies all homogeneous conditions and the necessary periodic conditions with respect to ϕ if

$$T''(t) + \lambda^2 a^2 T(t) = 0, \qquad T'(0) = 0$$

and if Φ and R are eigenfunctions of these two problems:

$$(5) \quad \Phi''(\phi) + \mu\Phi(\phi) = 0,\, \Phi(-\pi) = \Phi(\pi),\, \Phi'(-\pi) = \Phi'(\pi),$$
$$(6) \quad \rho^2 R''(\rho) + \rho R'(\rho) + (\lambda^2\rho^2 - \mu)R(\rho) = 0, \qquad R(c) = 0.$$

The eigenvalues of problem (5) are

$$\mu = n^2 \qquad\qquad (n = 0, 1, 2, \ldots),$$

and the eigenfunctions are $\cos n\phi$ and $\sin n\phi$, including $\Phi = 1$ when $n = 0$ (Sec. 34). Hence for each n, Φ can be any linear combination of $\cos n\phi$ and $\sin n\phi$. When $\mu = n^2$, the eigenvalues of problem (6), or

$$(\rho R')' + (\lambda^2\rho - n^2\rho^{-1})R = 0, \qquad R(c) = 0,$$

are the positive roots $\lambda = \lambda_{nj}$ of the equation

$$(7) \qquad\qquad J_n(\lambda c) = 0 \qquad (n = 0, 1, 2, \ldots).$$

The eigenfunctions are $R = J_n(\lambda_{nj}\rho)$ (Sec. 76). The function T is then $\cos(\lambda_{nj}at)$.

The generalized linear combination of our functions $R\Phi T$,

$$(8) \quad z = \sum_{n=0}^{\infty} \sum_{j=1}^{\infty} J_n(\lambda_{nj}\rho)(A_{nj}\cos n\phi + B_{nj}\sin n\phi)\cos(\lambda_{nj}at),$$

formally satisfies all homogeneous conditions. It satisfies the condition $z(\rho,\phi,0) = f(\rho,\phi)$ also, if the coefficients can be determined so that the series

$$(9) \quad \sum_{n=0}^{\infty}\left\{\left[\sum_{j=1}^{\infty} A_{nj}J_n(\lambda_{nj}\rho)\right]\cos n\phi + \left[\sum_{j=1}^{\infty} B_{nj}J_n(\lambda_{nj}\rho)\right]\sin n\phi\right\}$$

converges to $f(\rho,\phi)$ when $-\pi \leq \phi \leq \pi$, $0 \leq \rho \leq c$.

For each fixed value of ρ series (9) is the Fourier series for $f(\rho,\phi)$ on the interval $-\pi \leq \phi \leq \pi$ if the coefficients of $\cos n\phi$ and $\sin n\phi$ are the Fourier coefficients; that is, if

$$\sum_{j=1}^{\infty} A_{nj}J_n(\lambda_{nj}\rho) = \frac{1}{\pi}\int_{-\pi}^{\pi} f(\rho,\phi)\cos n\phi\, d\phi \qquad (n = 1, 2, \ldots),$$

$$= \frac{1}{2\pi}\int_{-\pi}^{\pi} f(\rho,\phi)\, d\phi \qquad\qquad (n = 0),$$

$$\sum_{j=1}^{\infty} B_{nj}J_n(\lambda_{nj}\rho) = \frac{1}{\pi}\int_{-\pi}^{\pi} f(\rho,\phi)\sin n\phi\, d\phi \qquad (n = 1, 2, \ldots).$$

For each fixed n the series here become Fourier-Bessel series representations of the functions of ρ represented by the right-hand members of the equations, on the interval $(0,c)$, provided that

(10) $\quad A_{nj} = \dfrac{2}{\pi c^2 [J_{n+1}(\lambda_{nj}c)]^2} \displaystyle\int_0^c \rho J_n(\lambda_{nj}\rho) \int_{-\pi}^{\pi} f(\rho,\phi) \cos n\phi \, d\phi \, d\rho,$

when $n = 1, 2, \ldots$, and that

(11) $\quad A_{0j} = \dfrac{1}{\pi c^2 [J_1(\lambda_{0j}c)]^2} \displaystyle\int_0^c \rho J_0(\lambda_{0j}\rho) \int_{-\pi}^{\pi} f(\rho,\phi) \, d\phi \, d\rho,$

(12) $\quad B_{nj} = \dfrac{2}{\pi c^2 [J_{n+1}(\lambda_{nj}c)]^2} \displaystyle\int_0^c \rho J_n(\lambda_{nj}\rho) \int_{-\pi}^{\pi} f(\rho,\phi) \sin n\phi \, d\phi \, d\rho.$

The displacements $z(\rho,\phi,t)$ are then given by formula (8) when the coefficients have the values given by equations (10), (11), and (12), assuming that the function f is such that the series in formula (8) has adequate properties of convergence and differentiability.

PROBLEMS

1. In the above problem suppose that the initial displacement function $f(\rho,\phi)$ is a linear combination of a finite number of the functions $J_n(\lambda_{nj}\rho) \cos n\phi$ and $J_n(\lambda_{nj}\rho) \sin n\phi$. Point out why the iterated series in formula (8) then contains only a finite number of terms and represents a rigorous solution of the boundary value problem.

2. In case the initial displacement of the membrane in Sec. 81 is $f(\rho)$, a function of ρ only, derive the formula

$$z(\rho,t) = \frac{2}{c^2} \sum_{j=1}^{\infty} \frac{J_0(\lambda_j\rho) \cos (\lambda_j at)}{[J_1(\lambda_j c)]^2} \int_0^c \xi J_0(\lambda_j \xi) f(\xi) \, d\xi,$$

where λ_j are the positive roots of $J_0(\lambda c) = 0$.

3. If the initial displacement of the membrane in Sec. 81 is $A J_0(\lambda_k\rho)$, where λ_k is some root of $J_0(\lambda c) = 0$, show that

$$z(\rho,t) = A J_0(\lambda_k\rho) \cos (\lambda_k at).$$

Observe that these displacements are all periodic in t with a common period; thus the membrane gives a musical note.

4. Replace the initial conditions (3), Sec. 81, by the conditions $z = 0$ and $z_t = 1$ when $t = 0$. This is the case if the membrane and its frame are moving with unit velocity in the z direction and the frame is brought to rest at the instant $t = 0$.   Derive the formula

$$z(\rho,t) = \frac{2}{ac} \sum_{j=1}^{\infty} \frac{\sin (\lambda_j at)}{\lambda_j^2 J_1(\lambda_j c)} J_0(\lambda_j \rho)$$

for the displacements, where λ_j are the positive roots of $J_0(\lambda c) = 0$.

5. Derive the following formula for the temperatures $u(\rho,\phi,t)$ in an infinite cylinder $\rho \leqq c$ if $u = 0$ on the surface $\rho = c$ and $u = f(\rho,\phi)$ when $t = 0$:

$$u = \sum_{n=0}^{\infty} \sum_{j=1}^{\infty} J_n(\lambda_{nj}\rho)(A_{nj} \cos n\phi + B_{nj} \sin n\phi) \exp (-\lambda_{nj}^2 kt),$$

where λ_{nj}, A_{nj}, and B_{nj} are the numbers defined in Sec. 81.

6. Derive a formula for the temperatures $u(\rho,z,t)$ in a solid cylinder $\rho \leqq c$, $0 \leqq z \leqq \pi$, whose entire surface is kept at temperature zero and whose initial temperature is A, a constant.   Show that the formula can be written

$$u(\rho,z,t) = A v(z,t)w(\rho,t),$$

where

$$v = \frac{4}{\pi} \sum_{n=1}^{\infty} \frac{\sin (2n - 1)z}{2n - 1} \exp [-(2n - 1)^2 kt]$$

and, when λ_j are the positive roots of $J_0(\lambda c) = 0$,

$$w = \frac{2}{c} \sum_{j=1}^{\infty} \frac{J_0(\lambda_j \rho)}{\lambda_j J_1(\lambda_j c)} \exp (-\lambda_j^2 kt).$$

Note that $v(z,t)$ represents temperatures in a slab $0 \leqq z \leqq \pi$ and $w(\rho,t)$ temperatures in an infinite cylinder $\rho \leqq c$, both with zero boundary temperature and unit initial temperature.

7. Derive the following formula for temperature $u(\rho,\phi,t)$ in the long right-angled cylindrical wedge bounded by the surface $\rho = 1$ and the planes $\phi = 0$ and $\phi = \frac{1}{2}\pi$ if $u = 0$ on those three surfaces and $u = f(\rho,\phi)$ when $t = 0$:

$$u = \sum_{n=1}^{\infty} \sum_{j=1}^{\infty} B_{nj} J_{2n}(\lambda_{nj}\rho) \sin (2n\phi) \exp (-\lambda_{nj}^2 kt),$$

where λ_{nj} are the positive roots of $J_{2n}(\lambda) = 0$ and

$$B_{nj}[J_{2n+1}(\lambda_{nj})]^2 \pi = 8 \int_0^{\pi/2} \sin 2n\phi \int_0^1 \rho J_{2n}(\lambda_{nj}\rho)f(\rho,\phi) \, d\rho \, d\phi.$$

8. If the face $\phi = \frac{1}{2}\pi$ of the wedge in Problem 7 is replaced by a face $\phi = \phi_0$, show why the formula for the temperatures will in general involve Bessel functions J_ν of *nonintegral* orders.

9. Solve Problem 7 if, instead of being kept at temperature zero, all three boundary surfaces of the wedge are insulated.

10. Solve the following problem for temperatures $u(\rho,t)$ in a thin circular plate with heat transfer from its faces into surroundings at temperature zero:

$$u_t = u_{\rho\rho} + \rho^{-1}u_\rho -- hu \quad (\rho < 1, t > 0; h > 0),$$
$$u(1,t) = 0, \qquad u(\rho,0) = 1.$$

Ans. $\quad u = 2e^{-ht} \sum_{j=1}^{\infty} \frac{J_0(\lambda_j\rho)}{\lambda_j J_1(\lambda_j)} \exp\left(-\lambda_j^2 t\right), \quad [J_0(\lambda_j) = 0, \lambda_j > 0].$

11. Solve Problem 10 after replacing the condition $u(1,t) = 0$ by this surface heat transfer condition at the edge:

$$u_\rho(1,t) = -h_0 u(1,t) \qquad\qquad (h_0 > 0).$$

12. Solve this Dirichlet problem for $V(\rho,z)$:

$$\begin{aligned} \nabla^2 V &= 0 & (\rho < 1, z > 0), \\ V(1,z) &= 0 & (z > 0), \\ V(\rho,0) &= 1 & (\rho < 1), \end{aligned}$$

and V is to be bounded in the domain $\rho < 1, z > 0$.

13. Let $u(\rho,z)$ denote steady temperatures in the semi-infinite cylinder $\rho \leqq 1, z \geqq 0$ whose base $z = 0$ is insulated, if $u(1,z) = 1$ when $0 < z < 1$ and $u(1,z) = 0$ when $z > 1$. Derive the formula

$$u(\rho,z) = \frac{2}{\pi} \int_0^\infty \frac{J_0(i\alpha\rho)}{\alpha J_0(i\alpha)} \cos \alpha z \sin \alpha \, d\alpha \qquad (i = \sqrt{-1}).$$

14. Given a function $f(z)$ that is represented by its Fourier integral formula for all real z, derive the following formula for the harmonic function $V(\rho,z)$ inside the cylinder $\rho = c$ such that $V(c,z) = f(z)$, $-\infty < z < \infty$:

$$V = \frac{1}{\pi} \int_0^\infty \frac{J_0(i\alpha\rho)}{J_0(i\alpha c)} \int_{-\infty}^\infty f(\xi) \cos [\alpha(z - \xi)] \, d\xi \, d\alpha.$$

CHAPTER 9

LEGENDRE POLYNOMIALS AND APPLICATIONS

82. Derivation of Legendre Polynomials. Separation of variables in Laplace's equation written in terms of the spherical coordinates r and θ, after writing x for $\cos \theta$, leads to *Legendre's equation*

$$(1) \qquad (1 - x^2)y''(x) - 2xy'(x) + \lambda y(x) = 0,$$

as we shall see later (Sec. 89). The points $x = 1$ and $x = -1$, corresponding to $\theta = 0$ and $\theta = \pi$, are singular points of that differential equation. We show now that there is a certain infinite set of values of the parameter λ for each of which equation (1) has a polynomial as a particular solution.

To determine whether Legendre's equation has a solution represented by a power series we substitute

$$(2) \qquad \qquad y = \sum_{j=0}^{\infty} a_j x^j$$

into equation (1). The substitution gives the equation

$$\sum_{j=0}^{\infty} [j(j - 1)x^{j-2}(1 - x^2) - 2jx^j + \lambda x^j]a_j = 0,$$

or, since $-j(j - 1) - 2j = -j(j + 1)$,

$$\sum_{j=0}^{\infty} [\lambda - j(j + 1)]a_j x^j + \sum_{k=2}^{\infty} k(k - 1)a_k x^{k-2} = 0.$$

In the last series we write $k = j + 2$; thus

$$(3) \qquad \sum_{j=0}^{\infty} \{[\lambda - j(j + 1)]a_j + (j + 2)(j + 1)a_{j+2}\}x^j = 0.$$

200

Equation (3) is an identity in x if the coefficients a_j, which are functions of λ, satisfy the recurrence relation

$$(4) \qquad a_{j+2} = -\frac{\lambda - j(j+1)}{(j+1)(j+2)}\, a_j \qquad (j = 0, 1, 2, \ldots).$$

The power series (2) then represents a solution of Legendre's equation within its interval of convergence provided its coefficients satisfy relation (4). This leaves a_0 and a_1 as arbitrary constants. Moreover if $a_0 = 0$, then $a_{2j} = 0$; while if $a_1 = 0$, then $a_{2j-1} = 0$, when $j = 1, 2, \ldots$.

But from relation (4) it is clear that when λ has any value of the set

$$(5) \qquad \lambda = n(n+1) \qquad (n = 0, 1, 2, \ldots),$$

then $a_{n+2} = 0$, and consequently $a_{n+4} = a_{n+6} = \cdots = 0$. If the integer n is odd and we take a_0 to be zero so that $a_{2j} = 0$ and if $a_1 \neq 0$, then the only nonvanishing coefficients are a_1, a_3, a_5, $\ldots$, a_n. That is, the series reduces to a polynomial of degree n containing only odd powers of x. If n is even and $a_1 = 0$ while $a_0 \neq 0$, the series reduces to a polynomial of degree n containing only even powers of x.

Thus *when* $\lambda = n(n+1)$, *there is always a polynomial solution of equation* (1). No questions of convergence or continuity arise.

If $n = 0$, the polynomial is a constant a_0; if $n = 1$, it is a_1x. When $n = 2, 3, \ldots$ and $\lambda = n(n+1)$, the recurrence relation (4) can be written

$$(6) \qquad a_{j+2} = -\frac{(n-j)(n+j+1)}{(j+1)(j+2)}\, a_j$$
$$(j = 0, 1, 2, \ldots, n-2).$$

Note that a_j, the coefficient of x^j, is a function $a_j(n)$ of n. We write $k = j + 2$ here and solve for a_{k-2}:

$$(7) \qquad a_{k-2} = -\frac{k(k-1)}{(n-k+2)(n+k-1)}\, a_k$$
$$(k = 2, 3, \ldots, n).$$

Whether n is even or odd, formula (7) gives the coefficients of the polynomial in terms of a_n since

$$a_{n-2} = -\frac{n(n-1)}{2(2n-1)}\, a_n,$$

$$a_{n-4} = -\frac{(n-2)(n-3)}{4(2n-3)}\, a_{n-2} = (-1)^2 \frac{n(n-1)(n-2)(n-3)}{2^2(2!)(2n-1)(2n-3)}\, a_n,$$

and so on.   That is,

$$(8) \quad a_{n-2j} = \frac{(-1)^j}{2^j j!} \frac{n(n-1) \cdots (n-2j+1)}{(2n-1)(2n-3) \cdots (2n-2j+1)} a_n.$$

Here a_n, that is, $a_n(n)$, is left arbitrary.   Let it be given this value:

$$a_n = \frac{(2n-1)(2n-3) \cdots (3)(1)}{n!}.$$

Then, after reducing the factorial expressions, we find that formula (8) can be written

$$(9) \quad a_{n-2j} = \frac{(-1)^j}{2^n j!} \frac{(2n-2j)!}{(n-2j)!(n-j)!}.$$

We use the convention that $0! = 1$.   If n is even, the polynomial begins with a constant term a_0 given by formula (9) when $j = \frac{1}{2}n$.   If n is odd, the term of lowest degree is $a_1 x$, where a_1 is given by formula (9) when $j = \frac{1}{2}(n-1)$.

A polynomial solution of Legendre's equation

$$(1 - x^2)y'' - 2xy' + n(n+1)y = 0 \qquad (n = 0, 1, 2, \ldots),$$

for all x, is therefore

$$y = P_n(x)$$

if $P_n(x)$ is the *Legendre polynomial* of degree n:

$$(10) \quad P_n(x) = \frac{1}{2^n} \sum_{j=0}^{m} \frac{(-1)^j}{j!} \frac{(2n-2j)!}{(n-2j)!(n-j)!} x^{n-2j},$$

where $m = \frac{1}{2}n$ if n is even or zero and $m = \frac{1}{2}(n-1)$ if n is odd. A simpler formula for P_n will be found in Sec. 84.

According to definition (10), the first few polynomials are

$$P_0(x) = 1, \quad P_1(x) = x, \quad P_2(x) = \tfrac{1}{2}(3x^2 - 1),$$
$$P_3(x) = \tfrac{1}{2}(5x^3 - 3x), \quad P_4(x) = \tfrac{1}{8}(35x^4 - 30x^2 + 3),$$
$$P_5(x) = \tfrac{1}{8}(63x^5 - 70x^3 + 15x).$$

83. Orthogonality.   The self-adjoint form of Legendre's equation is

$$(1) \quad \frac{d}{dx}\left[(1 - x^2) \frac{dy}{dx} \right] + \lambda y = 0.$$

It is a special case of the Sturm-Liouville equation (4), Sec. 32, in which $r(x) = 1 - x^2$, $p(x) = 1$, and $q(x) = 0$. Since $r(-1) = 0$ and $r(1) = 0$, no boundary conditions at the ends of the interval $-1 \leqq x \leqq 1$ are needed to complete the singular Sturm-Liouville problem on that interval; the problem consists of equation (1) and the condition that y and y' be continuous over that closed interval.

According to Theorem 3, Sec. 33, if two such continuous functions y_m and y_n satisfy equation (1) when $\lambda = \lambda_m$ and $\lambda = \lambda_n$, respectively, where $\lambda_m \neq \lambda_n$, then y_m is orthogonal to y_n on the interval $(-1,1)$ with weight function $p(x) = 1$. We have shown that the polynomial P_n satisfies equation (1) when $\lambda = n(n + 1)$. Since P_n and its derivatives are everywhere continuous it follows that

$$(2) \qquad \int_{-1}^{1} P_m(x)P_n(x)\, dx = 0 \qquad (m \neq n;\, m,\, n = 0,\, 1,\, 2,\, \ldots).$$

Our singular Sturm-Liouville problem on the interval $(-1,1)$ therefore has eigenvalues

$$(3) \qquad\qquad\qquad \lambda = n(n + 1) \qquad (n = 0,\, 1,\, 2,\, \ldots),$$

and the corresponding eigenfunctions

$$(4) \qquad\qquad\qquad y = P_n(x) \qquad (n = 0,\, 1,\, 2,\, \ldots)$$

satisfy the orthogonality property (2). In the notation used for inner products that property reads $(P_m, P_n) = 0$, $m \neq n$.

We shall prove (Secs. 87 and 88) that whenever a function f, together with its derivative f', is sectionally continuous on the interval $(-1,1)$, then, at each interior point x of the interval where f is continuous, the generalized Fourier series for f corresponding to the orthogonal set $\{P_n(x)\}$ converges to $f(x)$. Hence the set $\{P_n(x)\}$ is closed in the sense of pointwise convergence, and therefore complete in the class of all such functions f. It follows that *there can be no other eigenvalues than the set* (3). For suppose $y_0(x)$ is an eigenfunction (continuous, together with y_0' on the interval $-1 \leqq x \leqq 1$) corresponding to another eigenvalue λ_0. Then $(y_0, P_n) = 0$ for each n, and the completeness of the set $\{P_n\}$ requires that $y_0(x) = 0$ for all x on the open interval, so that y_0 cannot be an eigenfunction.

Since $P_n(x)$ is a polynomial containing only even powers of x if n is even and only odd powers if x is odd, it is an even or an odd function according as n is even or odd; that is,

(5) $$P_n(-x) = (-1)^n P_n(x) \quad (n = 0, 1, 2, \ldots).$$

Hence the product $P_m(x)P_n(x)$, in which m and n are both even or both odd, is an even function. It follows from property (2) that

(6) $$\int_0^1 P_{2m}(x)P_{2n}(x)\, dx = 0$$
$$(m \neq n; m, n = 0, 1, 2, \ldots),$$

(7) $$\int_0^1 P_{2m+1}(x)P_{2n+1}(x)\, dx = 0$$
$$(m \neq n; m, n = 0, 1, 2, \ldots).$$

Condition (6) states that *the set $\{P_{2n}\}$ of polynomials of even degree is orthogonal on the interval* $(0,1)$. Since $P'_{2n}(0) = 0$, those polynomials are the eigenfunctions of the singular Sturm-Liouville problem consisting of Legendre's equation (1) on the interval $(0,1)$ and the condition

(8) $$y'(0) = 0,$$

together with the condition that y and y' be continuous when $0 \leq x \leq 1$. The eigenvalues are $\lambda = 2n(2n + 1)$.

Similarly, according to condition (7), *the set $\{P_{2n+1}\}$ of polynomials of odd degree is orthogonal on the interval* $(0,1)$. Those polynomials are the eigenfunctions of the problem consisting of equation (1), the continuity conditions on the closed interval, and the boundary condition

(9) $$y(0) = 0.$$

The eigenvalues are $\lambda = (2n + 1)(2n + 2)$, where $n = 0, 1, 2, \ldots$.

The Fourier cosine series and sine series representations on half intervals followed from the basic Fourier series of sines and cosines. In just the same way, the representation of functions on the interval $(0,1)$ in terms of either P_{2n} or P_{2n+1} follows from the representation in terms of P_n on the interval $(-1,1)$.

Further formulas for $P_n(x)$ must be developed before we can establish the expansion theorem and certain order properties needed in the applications of the polynomials.[1] First, some remarks on other solutions of Legendre's equation.

---

[1] Readers who wish to pass to the applications at this time may read Theorem 1 (Sec. 86) and Sec. 87 and proceed to Sec. 89.

In the preceding section the polynomials P_n were found by setting one of the two arbitrary constants a_0 or a_1, in the series, equal to zero. If those constants are left arbitrary, the series furnishes the general solution of Legendre's equation, when $\lambda = n(n + 1)$, in the form

$$(10) \qquad\qquad y = AP_n(x) + BQ_n(x),$$

where Q_n is *Legendre's function of the second kind.* When $|x| < 1$, $Q_n(x)$ is represented by a power series in x whose coefficients are found with the aid of the recurrence relation (6), Sec. 82. But Q_n is not continuous at the end points of the interval $(-1,1)$. When λ has an arbitrary value, two fundamental solutions of Legendre's equation on the interval $-1 < x < 1$ can be written as power series in x by using the recurrence relation (4), Sec. 83.

84. Rodrigues' Formula. Norms. The result of integrating the polynomial $P_n(x)$ defined by equation (10), Sec. 82, n times from 0 to t is this polynomial $S_{2n}(t)$ of degree $2n$:

$$\frac{1}{2^n} \sum_{j=0}^{m} \frac{(-1)^j(2n - 2j)!}{j!(n - j)!(n - 2j)!} \times$$

$$\frac{t^{2n-2j}}{(n - 2j + 1)(n - 2j + 2)\cdots(2n - 2j)}.$$

In terms of binomial coefficients,

$$(1) \qquad S_{2n}(t) = \frac{1}{2^n n!} \sum_{j=0}^{m} (-1)^j \frac{n!}{j!(n - j)!} (t^2)^{n-j}.$$

The lowest power of t in $S_{2n}(t)$ is t^{2n-2m}, that is, t^n if n is even, or t^{n+1} if n is odd. If the sum in equation (1) is extended to the range $j = 0$ to $j = n$, it represents the binomial expansion of $(t^2 - 1)^n$. The additional polynomial that is introduced is of degree less than n, one whose nth derivative is zero. Since the nth derivative of $S_{2n}(t)$ is $P_n(t)$, it follows that $P_n(t)$ is also the nth derivative of $(2^n n!)^{-1}(t^2 - 1)^n$; that is,

$$(2) \qquad P_n(x) = \frac{1}{2^n n!} \frac{d^n}{dx^n} (x^2 - 1)^n \qquad (n = 0, 1, 2, \ldots).$$

This is *Rodrigues' formula* for the Legendre polynomials.

We write u for $x^2 - 1$.   Then

$$u^n = (x^2 - 1)^n = (x + 1)^n (x - 1)^n.$$

By Leibnitz's rule for the nth derivative D^n of a product,

$$D^n(u^n) = \sum_{k=0}^{n} \frac{n!}{k!(n-k)!} D^k[(x+1)^n] D^{n-k}[(x-1)^n],$$

where $D^0 w = w$.   Only the first term $(k = 0)$ in this sum is free of the factor $x - 1$.   Consequently when $x = 1$, the value of the sum is $n! 2^n$.   It follows from Rodrigues' formula that

$$(3) \qquad\qquad\qquad P_n(1) = 1 \qquad\quad (n = 0, 1, 2, \ldots).$$

A recurrence relation for P_n follows from Rodrigues' formula by first noting that

$$2^{n+1}(n + 1)! P_{n+1} = D^{n+1}(u^{n+1}) = D^{n-1}(D^2 u^{n+1}).$$

But $Du^{n+1} = 2(n + 1)xu^n$

and $\begin{aligned}[t] D^2 u^{n+1} &= 2(n + 1)(u^n + 2nx^2 u^{n-1}) \\ &= 2(n + 1)[u^n + 2n(x^2 - 1)u^{n-1} + 2nu^{n-1}] \\ &= 2(n + 1)[(2n + 1)u^n + 2nu^{n-1}]. \end{aligned}$

Consequently

$$2^n n! P_{n+1} = (2n + 1)D^{n-1}u^n + 2nD^{n-1}u^{n-1};$$

that is,

$$(4) \qquad\quad P_{n+1}(x) - P_{n-1}(x) = \frac{2n + 1}{2^n n!} D^{n-1}u^n.$$

With the aid of Leibnitz's rule we can write

$$\begin{aligned} P_{n+1} &= \frac{D^n(Du^{n+1})}{2^{n+1}(n + 1)!} = \frac{D^n(xu^n)}{2^n n!} \\ &= \frac{xD^n u^n + nD^{n-1}u^n}{2^n n!} = xP_n + \frac{n}{2^n n!} D^{n-1}u^n. \end{aligned}$$

Elimination of $D^{n-1}u^n$ between this equation and equation (4) gives the *recurrence relation*

$$(5) \qquad (n + 1)P_{n+1}(x) + nP_{n-1}(x) = (2n + 1)xP_n(x)$$
$$(n = 1, 2, \ldots).$$

As an immediate consequence of equation (4),

(6) $P'_{n+1}(x) - P'_{n-1}(x) = (2n + 1)P_n(x)$ $(n = 1, 2, \ldots)$.

Relation (5), written in the form

(7) $nP_n(x) = (2n - 1)xP_{n-1}(x) - (n - 1)P_{n-2}(x)$

$$(n = 2, 3, \ldots),$$

can be used to find norms of the orthogonal polynomials P_n on the interval $(-1,1)$. Since $(P_n, P_{n-2}) = 0$ and the integrals representing (P_n, xP_{n-1}) and (xP_n, P_{n-1}) are identical, we find from formulas (7) and (5) that

$$(P_n, P_n) = \frac{2n - 1}{n} (xP_n, P_{n-1}) = \frac{2n - 1}{2n + 1} (P_{n-1}, P_{n-1}).$$

Continuing that reduction formula,

$$(P_n, P_n) = \frac{2n - 1}{2n + 1} \frac{2n - 3}{2n - 1} (P_{n-2}, P_{n-2}) = \frac{2n - 5}{2n + 1} (P_{n-3}, P_{n-3})$$
$$= \cdots = \frac{2n - (2n - 1)}{2n + 1} (P_0, P_0) = \frac{2}{2n + 1}.$$

The result is easily verified in case $n = 0$ or $n = 1$, Thus

(8) $\|P_n\|^2 = \displaystyle\int_{-1}^{1} [P_n(x)]^2 \, dx = \frac{2}{2n + 1}$ $(n = 0, 1, 2, \ldots)$.

PROBLEMS

1. Establish these properties of $P_n(x)$:

(a) $P_n(-1) = (-1)^n$;   (b) $P_{2n+1}(0) = 0$;

(c) $P_{2n}(0) = (-1)^n \dfrac{(2n)!}{2^{2n}(n!)^2}$;   (d) $P'_{2n}(0) = 0$.

2. Draw the graphs of $P_0(x)$, $P_1(x)$, and $P_2(x)$.

3. Verify directly that the set of three polynomials, $P_0(x)$, $P_1(x)$, $P_2(x)$, is orthogonal on the interval $(-1,1)$.

4. In Sec. 82, give the details in reducing formula (8) to the form (9) when a_n has the value given there.

5. From formula (6), Sec. 84, obtain the integration formula

$$\int_{x}^{1} P_n(\xi) \, d\xi = \frac{1}{2n + 1} [P_{n-1}(x) - P_{n+1}(x)]   (n = 1, 2, \ldots).$$

6. From the orthogonality of the set $\{P_n\}$ state why

(a) $\displaystyle\int_{-1}^{1} P_n(x)\, dx = 0$ when $n = 1, 2, \ldots$;

(b) $\displaystyle\int_{-1}^{1} (Ax + B)P_n(x)\, dx = 0$ when $n = 2, 3, \ldots$ (A, B constant).

7. Why is the set of functions $\{\sqrt{4n+1}\, P_{2n}(x)\}$, $n = 0, 1, 2, \ldots$ *orthonormal on the interval* $(0,1)$?

8. Why is the set of functions $\{\sqrt{4n+3}\, P_{2n+1}(x)\}$, $n = 0, 1, 2, \ldots$ *orthonormal on the interval* $(0,1)$?

85. Integral Forms. Useful orders of magnitude of $P_n(x)$, with respect to n or x, follow from certain representations of those polynomials by integrals.

In Sec. 31 we noted that

$$\int_{-\pi}^{\pi} e^{im\phi}\, d\phi = 0 \qquad \text{when } m = 1, 2, \ldots$$

and the value of the integral is 2π when $m = 0$. If $r = r(x)$, where the variable x is independent of ϕ, then from the nature of the binomial expansion of $(x + re^{i\phi})^n$ it follows that

$$\frac{1}{2\pi} \int_{-\pi}^{\pi} (x + re^{i\phi})^n\, d\phi = x^n \qquad (n = 0, 1, 2, \ldots).$$

Consequently if $p(x)$ is any polynomial in x, then

$$(1) \qquad p(x) = \frac{1}{2\pi} \int_{-\pi}^{\pi} p(x + re^{i\phi})\, d\phi.$$

Each derivative of $p(x)$ is also a polynomial. Thus we find, after an integration by parts, that

$$2\pi p^{(n)}(x) = \int_{-\pi}^{\pi} p^{(n)}(x + re^{i\phi})\, d\phi$$

$$= \frac{1}{r} \int_{-\pi}^{\pi} e^{-i\phi} p^{(n-1)}(x + re^{i\phi})\, d\phi \quad [r(x) \neq 0].$$

Upon continuing the integration by parts we find that

$$(2) \qquad p^{(n)}(x) = \frac{n!}{2\pi r^n} \int_{-\pi}^{\pi} e^{-in\phi} p(x + re^{i\phi})\, d\phi$$

whenever $r(x) \neq 0$. Details here are left to the problems.

We now define p and r as follows;

(3) $\qquad p(x) = \dfrac{1}{2^n n!} (x^2 - 1)^n, \qquad r(x) = i(1 - x^2)^{\frac{1}{2}}.$

Elementary simplifications then show that

(4) $\qquad n! p(x + re^{i\phi}) = r^n e^{in\phi}(x + r \cos \phi)^n.$

Rodrigues' formula now reads $P_n(x) = p^{(n)}(x)$. According to equations (2) and (4) therefore, P_n has the *integral representation*

(5) $\qquad \begin{aligned} P_n(x) &= \frac{1}{2\pi} \int_{-\pi}^{\pi} (x + i \sqrt{1 - x^2} \cos \phi)^n \, d\phi \\ &= \frac{1}{\pi} \int_0^{\pi} (x + i \sqrt{1 - x^2} \cos \phi)^n \, d\phi. \end{aligned}$

Note that the formula is correct when $x = \pm 1$, and therefore for all x.

When $x = \cos \theta$, formula (5) takes the form

(6) $\qquad P_n(\cos \theta) = \frac{1}{\pi} \int_0^{\pi} (\cos \theta + i \sin \theta \cos \phi)^n \, d\phi$

$\qquad\qquad\qquad\qquad\qquad\qquad\qquad (0 \le \theta \le \pi).$

By using $\pi - \phi$ as the variable of integration over the half interval $(\frac{1}{2}\pi, \pi)$ we can write the integral here as

$$\int_0^{\pi/2} [(\cos \theta + i \sin \theta \cos \phi)^n + (\cos \theta - i \sin \theta \cos \phi)^n] \, d\phi.$$

Since $(\bar{z})^n$ is the complex conjugate of z^n, the integrand is twice the real part of its first term; thus

(7) $\qquad P_n(\cos \theta) = \frac{2}{\pi} \int_0^{\pi/2} \Re[(\cos \theta + i \sin \theta \cos \phi)^n] \, d\phi$

$\qquad\qquad\qquad\qquad\qquad\qquad\qquad (0 \le \theta \le \pi).$

Furthermore $|\Re(z)| \le |z|$; therefore

$$|P_n(\cos \theta)| \le \frac{2}{\pi} \int_0^{\pi/2} (\cos^2 \theta + \sin^2 \theta \cos^2 \phi)^{n/2} \, d\phi;$$

that is,

(8) $\qquad |P_n(\cos \theta)| \le \frac{2}{\pi} \int_0^{\pi/2} (1 - \sin^2 \theta \sin^2 \phi)^{n/2} \, d\phi.$

From condition (8) it follows that $|P_n(\cos \theta)| \leqq 1$ or

$$(9) \qquad |P_n(x)| \leqq 1 \qquad \text{when } -1 \leqq x \leqq 1$$
$$(n = 0, 1, 2, \ldots).$$

The graph of $\sin \phi$ shows that $\sin \phi > 2\phi/\pi$ when $0 < \phi < \frac{1}{2}\pi$. Thus if $y = (2\phi/\pi)^2 \sin^2 \theta$, it follows that

$$1 - \sin^2 \theta \sin^2 \phi < 1 - y \quad \text{when } 0 < \phi < \frac{1}{2}\pi.$$

Also, $1 - y \leqq e^{-y}$, as can be seen graphically or from Maclaurin's series for e^{-y}. According to condition (8) therefore,

$$|P_n(\cos \theta)| < \frac{2}{\pi} \int_0^{\pi/2} \exp\left(-\frac{2n \sin^2\theta}{\pi^2} \phi^2\right) d\phi.$$

We substitute $t = \phi\pi^{-1} \sqrt{2n} \sin \theta$ to write

$$|P_n(\cos \theta)| < \left(\frac{2}{n}\right)^{\frac{1}{2}} \frac{1}{\sin \theta} \int_0^\infty \exp\left(-t^2\right) dt = \left(\frac{\pi}{2n}\right)^{\frac{1}{2}} \frac{1}{\sin \theta}.$$

This gives the following order property of P_n, with respect to n. *For each fixed x $(-1 < x < 1)$, $P_n(x)$ is of the order of $n^{-\frac{1}{2}}$ as n increases:*

$$(10) \qquad \sqrt{n} \, |P_n(x)| < \left[\frac{\pi}{2(1 - x^2)}\right]^{\frac{1}{2}}$$
$$(-1 < x < 1, n = 1, 2, \ldots).$$

86. Further Order Properties. Since $P_n(x)$ is a polynomial of degree n containing only alternate powers of x, then

$$x^n = aP_n(x) + bx^{n-2} + cx^{n-4} + \cdots,$$

where the coefficients are constants. Similarly, x^{n-2} is a linear combination of $P_{n-2}(x)$ and a polynomial of degree $n - 4$, and so on. Thus x^n is a linear combination of the polynomials $P_k(x)$, where $k = n, n - 2, \ldots, 1$ or 0.

The derivative $P_n'(x)$ is a polynomial of degree $n - 1$, containing alternate powers of x, each of which can be written as a linear combination of Legendre polynomials. Therefore P_n' can be written as the following linear combination:

$$(1) \qquad P_n'(x) = A_{n-1}P_{n-1}(x) + A_{n-3}P_{n-3}(x) + \cdots.$$

To find A_j we take inner products of both members with P_j,

$$\int_{-1}^{1} P_j(x)P'_n(x)\,dx = A_j(P_j,P_j) \quad (j = n - 1, n - 3, \ldots).$$

When integrated by parts, the integral on the left becomes

$$P_j(x)P_n(x)\Big]_{-1}^{1} - \int_{-1}^{1} P_n(x)P'_j(x)\,dx,$$

and this last integral vanishes because P'_j is a linear combination of Legendre polynomials of degree less than n. But $P_k(1) = 1$, $P_k(-1) = (-1)^k$, and $(P_j,P_j) = 2(2j + 1)^{-1}$. Therefore

$$A_j = 2j + 1 \quad (j = n - 1, n - 3, \ldots).$$

The representation (1), valid for all x, then becomes

$$(2) \qquad P'_n(x) = (2n - 1)P_{n-1}(x) + (2n - 5)P_{n-3}(x) + \ldots$$
$$(n = 1, 2, \ldots),$$

ending with $3P_1(x)$ if n is even and with $P_0(x)$ if n is odd.

Assume now that $-1 \leqq x \leqq 1$. Then $|P_n(x)| \leqq 1$ (Sec. 85), and it follows from formula (2) that

$$|P'_{2n}(x)| \leqq (4n - 1) + (4n - 5) + \cdots + 3 = n(2n + 1),$$
$$|P'_{2n+1}(x)| \leqq (4n + 1) + (4n - 3) + \cdots + 1$$
$$= (n + 1)(2n + 1).$$

Since $n(2n + 1) \leqq (2n)^2$ and $(n + 1)(2n + 1) \leqq (2n + 1)^2$, then

$$(3) \qquad\qquad |P'_n(x)| \leqq n^2 \qquad \text{when } -1 \leqq x \leqq 1$$
$$(n = 0, 1, 2, \ldots).$$

Differentiating both members of equation (2) and noting that $|P'_{n-1}(x)| < n^2$, $|P'_{n-3}(x)| < n^2$, etc., when $|x| \leqq 1$, we see by the method used above that

$$(4) \qquad\qquad |P''_n(x)| \leqq n^4 \qquad \text{when } -1 \leqq x \leqq 1$$
$$(n = 0, 1, 2, \ldots).$$

Similarly, $|P_n^{(k)}(x)| \leqq n^{2k}$ when $k = 3, 4, \ldots$ and $|x| \leqq 1$.

We collect our order properties of P_n as follows.

Theorem 1. *For each positive integer n, at all points of the interval $-1 \leqq x \leqq 1$, the values of each of the functions*

$$|P_n(x)|, \frac{1}{n^2}\,|P'_n(x)|, \frac{1}{n^4}\,|P''_n(x)|, \cdots$$

never exceed unity.   For each fixed x_0 $(-1 < x_0 < 1)$.

$$|P_n(x_0)| < \frac{M_0}{\sqrt{n}} \qquad (n = 1, 2, \ldots),$$

where (Sec. 85) the value of M_0 depends only on x_0.

87. Legendre's Series. From the orthogonality of P_n and formula (8), Sec. 84, for the norms of those polynomials it follows that the set of functions

(1) $$\phi_n(x) = \sqrt{n + \tfrac{1}{2}} \, P_n(x) \quad (n = 0, 1, 2, \ldots)$$

is orthonormal on the interval $(-1,1)$. The Fourier constants with respect to that set, of a function f defined on the interval, are

$$c_n = (f, \phi_n) = \sqrt{n + \tfrac{1}{2}} \int_{-1}^{1} f(\xi) P_n(\xi) \, d\xi.$$

The generalized Fourier series corresponding to f is

$$\sum_{n=0}^{\infty} c_n \phi_n(x) = \sum_{n=0}^{\infty} (n + \tfrac{1}{2}) P_n(x) \int_{-1}^{1} f(\xi) P_n(\xi) \, d\xi.$$

That correspondence can be written

(2) $$\sum_{n=0}^{\infty} A_n P_n(x) \sim f(x) \qquad (-1 < x < 1),$$

where

(3) $$A_n = \frac{2n + 1}{2} \int_{-1}^{1} f(x) P_n(x) \, dx \qquad (n = 0, 1, 2, \ldots).$$

Series (2) with coefficients (3) is *Legendre's series* corresponding to the function f. In the following section we shall prove that it converges to $f(x)$ under conditions stated in this theorem:

Theorem 2. *Let f denote a sectionally continuous function on the interval $(-1,1)$. Then at each interior point x of the interval at which f is continuous and has derivatives from the right and left, Legendre's series corresponding to f converges to $f(x)$; that is,*

(4) $$f(x) = \sum_{n=0}^{\infty} A_n P_n(x) \qquad (-1 < x < 1),$$

where the coefficients A_n are given by formula (3).

The proof can be extended to show that the series converges to the mean of the values $f(x + 0)$ and $f(x - 0)$ when f has a jump at the interior point x, if both one-sided derivatives exist there.[1] That result will not be needed here.

In case f is an *even function*, then $f(x)P_n(x)$ is even when n is even, and odd when n is odd. Hence $A_{2n+1} = 0$ and

$$(5) \quad A_{2n} = (4n + 1) \int_0^1 f(x)P_{2n}(x)\, dx \qquad (n = 0, 1, 2, \ldots).$$

Thus *when f is sectionally continuous on the interval $(0,1)$ and its one-sided derivatives exist at a point x $(0 < x < 1)$ at which f is continuous and when A_{2n} have the values (5), then*

$$(6) \qquad\qquad f(x) = \sum_{n=0}^{\infty} A_{2n}P_{2n}(x) \qquad (0 < x < 1)$$

because Theorem 2 applies to the even extension of f.

Similarly, *when f satisfies those same conditions on the interval $(0,1)$ and at the point x, we find that*

$$(7) \qquad\qquad f(x) = \sum_{n=0}^{\infty} A_{2n+1}P_{2n+1}(x) \qquad (0 < x < 1)$$

where

$$(8) \quad A_{2n+1} = (4n + 3) \int_0^1 f(x)P_{2n+1}\, dx \qquad (n = 0, 1, 2, \ldots).$$

Each of the two sets of polynomials $\{P_{2n}\}$ and $\{P_{2n+1}\}$, orthogonal on the interval $(0,1)$, is therefore closed in the sense of pointwise convergence.

88. Convergence of the Series. To prove Theorem 2, let x_0 denote an interior point of the interval $(-1,1)$ at which the function f is continuous and has one-sided derivatives. At x_0 the value of the sum of the first m terms of Legendre's series can be written

$$(1) \qquad S_m(x_0) = \sum_{n=0}^{m} A_n P_n(x_0) = \int_{-1}^{1} f(x)K_m(x,x_0)\, dx$$

[1] See D. Jackson, "Fourier Series and Orthogonal Polynomials," pp. 65ff.

where $-1 < x_0 < 1$ and

$$(2) \qquad K_m(x,x_0) = \sum_{n=0}^{m} \frac{2n+1}{2} P_n(x)P_n(x_0).$$

According to the recurrence formula (5), Sec. 84,

$$(3) \qquad (2n+1)xP_n(x) = (n+1)P_{n+1}(x) + nP_{n-1}(x)$$
$$(n = 1, 2, \ldots).$$

This leads to a more compact expression for K_m. After multiplying the members of equation (3) by $P_n(x_0)$ and the members of the equation

$$(2n+1)x_0P_n(x_0) = (n+1)P_{n+1}(x_0) + nP_{n-1}(x_0)$$

by $P_n(x)$ and subtracting, and introducing the notation

$$(4) \qquad R_n(x,x_0) = P_n(x)P_{n-1}(x_0) - P_{n-1}(x)P_n(x_0),$$

we find that

$$(2n+1)(x - x_0)P_n(x)P_n(x_0) = (n+1)R_{n+1}(x,x_0) - nR_n(x,x_0).$$

When the expression here for $(2n+1)P_n(x)P_n(x_0)$ is used in the sum (2), terms of the sum cancel by pairs, giving

$$(5) \qquad K_m(x,x_0) = \frac{m+1}{2}\frac{P_{m+1}(x)P_m(x_0) - P_m(x)P_{m+1}(x_0)}{x - x_0}$$

if $x \neq x_0$. But the numerator of the last fraction, which is $R_{m+1}(x,x_0)$, is a polynomial in x that vanishes when $x = x_0$, so it contains $x - x_0$ as a factor; therefore the fraction has a limit as $x \to x_0$. Note that equation (5) agrees with definition (2) when $m = 0$.

According to equations (1), (4), and (5),

$$(6) \qquad S_m(x_0) = \frac{m+1}{2}\int_{-1}^{1} f(x)\frac{R_{m+1}(x,x_0)}{x - x_0}\,dx.$$

Consider for the moment the special case $f(x) = 1 = P_0(x)$. Then $A_0 = 1$ and $A_n = 0$ $(n = 1, 2, \ldots)$ and, in view of equation (1), $S_m(x_0) = 1$. Thus from eqution (6),

$$(7) \qquad 1 = \frac{m+1}{2}\int_{-1}^{1}\frac{R_{m+1}(x,x_0)}{x - x_0}\,dx \qquad (-1 < x_0 < 1).$$

We now multiply the members of equation (7) by $f(x_0)$ and subtract from the members of equation (6) to write

$$S_m(x_0) - f(x_0) = \frac{m+1}{2} \int_{-1}^{1} \frac{f(x) - f(x_0)}{x - x_0} R_{m+1}(x,x_0) \, dx;$$

that is,

$$(8) \quad S_m(x_0) - f(x_0)$$
$$= \frac{m+1}{2} \int_{-1}^{1} F_0(x)[P_{m+1}(x)P_m(x_0) - P_m(x)P_{m+1}(x_0)] \, dx,$$

where $$F_0(x) = \frac{f(x) - f(x_0)}{x - x_0}.$$

Since $F_0(x_0 + 0)$ and $F_0(x_0 - 0)$ are the one-sided derivatives of the given function f at the point x_0, F_0 is sectionally continuous on the interval $(-1,1)$. Let C_m denote the Fourier constants of F_0 corresponding to the orthonormal set of functions $\phi_m = (m + \frac{1}{2})^{\frac{1}{2}} P_m$. Then equation (8) can be written

$$(9) \quad S_m(x_0) - f(x_0)$$
$$= \frac{m+1}{2(m+\frac{3}{2})^{\frac{1}{2}}} P_m(x_0)C_{m+1} - \frac{m+1}{2(m+\frac{1}{2})^{\frac{1}{2}}} P_{m+1}(x_0)C_m.$$

But $|P_m(x_0)| < M_0 m^{-\frac{1}{2}}$, where M_0 is independent of m (Theorem 1). Thus it follows from equation (9) that

$$|S_m(x_0) - f(x_0)| < M_0 \frac{m+1}{2m} (|C_{m+1}| + |C_m|)$$

and, since $C_m \to 0$ as $m \to \infty$ (Sec. 28), it is true that

$$(10) \qquad\qquad \lim_{m \to \infty} S_m(x_0) = f(x_0) \qquad (-1 < x_0 < 1).$$

This completes the proof of Theorem 2.

PROBLEMS

1. If $f(x) = 0$ when $-1 < x < 0$ and $f(x) = 1$ when $0 < x < 1$, use Theorem 2 and results found in Problems 1 and 5, Sec. 84, to show that

$$f(x) = \frac{1}{2} P_0(x) + \frac{1}{2} \sum_{n=0}^{\infty} [P_{2n}(0) - P_{2n+2}(0)]P_{2n+1}(x)$$

$$= \frac{1}{2} + \frac{3}{4} x + \sum_{n=1}^{\infty} (-1)^n \frac{4n+3}{4n+4} \frac{(2n)!}{2^{2n}(n!)^2} P_{2n+1}(x)$$

when $-1 < x < 1$ and $x \neq 0$. Note that if $f(0)$ is defined to be $\frac{1}{2}$ the expansion is also valid when $x = 0$.

2. If $f(x) = 0$ when $-1 < x \leq 0$ and $f(x) = x$ when $0 \leq x < 1$, (a) state why f is represented by its Legendre series at each point of the interval $-1 < x < 1$; (b) show that $A_{2n+1} = 0$ when $n = 1, 2, \ldots$; (c) find the first four nonvanishing terms of the expansion.

Ans. $f(x) = \frac{1}{4} + \frac{1}{2}P_1(x) + \frac{5}{16}P_2(x) - \frac{3}{32}P_4(x) + \cdots$

$$(-1 < x < 1).$$

3. Prove that, for all x,

(a) $x^2 = \frac{1}{3}P_0(x) + \frac{2}{3}P_2(x)$; (b) $x^3 = \frac{3}{5}P_1(x) + \frac{2}{5}P_3(x)$.

4. Expand the function $f(x) = 1$ $(0 < x < 1)$ in series of Legendre polynomials of odd degree, on the interval $(0,1)$. What function does the series represent on the interval $-1 < x < 0$?

$$Ans. \quad 1 = \sum_{n=0}^{\infty} (-1)^n \frac{4n+3}{2n+2} \frac{(2n)!}{2^{2n}(n!)^2} P_{2n+1}(x) \quad (0 < x < 1).$$

5. Obtain the first few terms in the representation of the function $f(x) = x$ over the interval $0 \leq x < 1$ in series of Legendre polynomials of even degree to show that

$$x = \frac{1}{2}P_0(x) + \frac{5}{8}P_2(x) - \frac{3}{16}P_4(x) + \cdots \quad (0 \leq x < 1).$$

Note why the representation must be valid when $x = 0$ and what function the series represents on the interval $-1 < x < 1$.

6. State why it is true that (a) when $-1 < x_0 < 1$, then

$$\lim_{n \to \infty} P_n(x_0) = 0 \qquad (n = 0, 1, 2, \ldots);$$

(b) when f is sectionally continuous on the interval $(0,1)$, then

$$\lim_{n \to \infty} (4n + 1)^{\frac{1}{2}} \int_0^1 f(x)P_{2n}(x) \, dx = 0.$$

7. Give details in the derivation of equation (2), Sec. 85.

8. Give details in the derivation of formula (5), Sec. 88.

89. Dirichlet Problems in Spherical Regions. As our first application of Legendre's series we determine the harmonic function V in the domain $r < c$ such that V assumes prescribed values $F(\theta)$ on the spherical surface $r = c$ (Fig. 24). Here r, ϕ, and θ are spherical coordinates, and V is independent of ϕ. Thus

V satisfies Laplace's equation

(1)
$$r \frac{\partial^2}{\partial r^2}(rV) + \frac{1}{\sin\theta}\frac{\partial}{\partial\theta}\left(\sin\theta\frac{\partial V}{\partial\theta}\right) = 0$$

in the domain $r < c$, $0 < \theta < \pi$ and the condition

(2)
$$\lim_{r\to c} V = F(\theta) \qquad (0 < \theta < \pi, r < c);$$

also V and its partial derivatives of first and second order are to be continuous throughout the interior $(0 \leqq r < c, 0 \leqq \theta \leqq \pi)$ of the sphere.

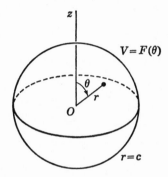

FIG. 24

Physically, the function V may denote steady temperatures in a solid sphere $r \leqq c$ whose surface temperature depends only on θ; that is, the surface temperature is uniform over each circle $\theta = \theta_0$, $r = c$. Also, V represents electrostatic potential in the space $r < c$ free of charges, when $V = F(\theta)$ on the boundary $r = c$.

Let us introduce a new variable x here, where

$$x = \cos\theta \qquad (0 \leqq \theta \leqq \pi).$$

In equation (1) the differential form in θ can be written

$$\frac{1}{\sin\theta}\frac{\partial}{\partial\theta}\left[(1 - \cos^2\theta)\frac{1}{\sin\theta}\frac{\partial V}{\partial\theta}\right] = \frac{\partial}{\partial x}\left[(1 - x^2)\frac{\partial V}{\partial x}\right].$$

If we write $F(\theta) = f(\cos\theta)$, then $V(r,x)$ satisfies the conditions

(3) $r(rV)_{rr} + [(1 - x^2)V_x]_x = 0$ $\qquad (r < c, -1 < x < 1),$

(4) $\qquad\qquad V(c - 0, x) = f(x)$ $\qquad (-1 < x < 1),$

and V and its derivatives of first and second order are to be continuous functions of r and x when $0 \leqq r < c$, $-1 \leqq x \leqq 1$.

Consider a solution of equation (3) of the form $R(r)X(x)$ that satisfies the continuity requirements. Separation of variables shows that, for some constant λ, $X(x)$ must satisfy Legendre's equation

$$(5) \qquad [(1 - x^2)X']' + \lambda X = 0 \qquad (-1 < x < 1),$$

where X, X' and X'' are to be continuous over the closed interval $-1 \leqq x \leqq 1$. Except when $X(x)$ is identically zero, we have seen (Sec. 83) that those conditions are satisfied only in case λ has one of these eigenvalues:

$$\lambda = n(n + 1) \qquad (n = 0, 1, 2, \ldots);$$

then $X = P_n(x)$. The separation also requires that, for the same constant λ, $r(rR)'' = \lambda R$; that is,

$$(6) \qquad r^2R'' + 2rR' - n(n + 1)R = 0 \qquad (r < 1).$$

Also, R is to be continuous when $0 \leqq r < 1$.

Equation (6) has the form of Cauchy's linear equation, which reduces to one with constant coefficients after the substitution $r = e^t$. Its general solution is

$$(7) \qquad R = C_1 r^n + C_2 r^{-n-1},$$

as is readily verified. The continuity of R when $0 \leqq r < c$, at the origin $r = 0$ in particular, requires that $C_2 = 0$.

The functions $r^n P_n(x)$ therefore satisfy Laplace's equation (3) and the continuity conditions. Formally, their generalized linear combination

$$(8) \qquad V(r,x) = \sum_{n=0}^{\infty} A_n \left(\frac{r}{c}\right)^n P_n(x) \qquad (r \leqq c)$$

is a solution of our boundary value problem if the coefficients A_n are such that $V(c,x) = f(x)$; that is,

$$(9) \qquad f(x) = \sum_{n=0}^{\infty} A_n P_n(x) \qquad (-1 < x < 1).$$

We assume that f and f' are sectionally continuous over the interval $(-1,1)$. Then equation (9) is Legendre's series representation of f, valid at each point $(-1 < x < 1)$ where f is continuous (Theorem 2), when

$$(10) \quad A_n = \frac{2n + 1}{2} \int_{-1}^{1} f(\xi) P_n(\xi) \, d\xi \qquad\qquad (n = 0, 1, 2, \ldots).$$

The harmonic function V is then given by formula (8) with the coefficients (10). As a function of r and θ,

$$(11) \quad V = \sum_{n=0}^{\infty} \frac{2n + 1}{2} \left(\frac{r}{c}\right)^n P_n(\cos \theta) \int_{-1}^{1} f(\xi) P_n(\xi) \, d\xi \qquad (r \leqq c).$$

A full verification of this solution is given below.

The harmonic function W in the unbounded domain $r > c$, exterior to the spherical surface $r = c$, which assumes the values $f(\cos \theta)$ on that surface and which tends to zero as $r \to \infty$, can be found in like manner. Here $C_1 = 0$ in equation (7) if R is to vanish as $r \to \infty$, and our particular solutions are $r^{-n-1} P_n(x)$. The formula for W then becomes

$$(12) \qquad\qquad W(r,x) = \sum_{n=0}^{\infty} A_n \left(\frac{c}{r}\right)^{n+1} P_n(x) \qquad\qquad (r \geqq c),$$

where A_n have the values (10) and $x = \cos \theta$.

Verification. Formula (8) can be established as a solution of our problem in V by the method used in Sec. 61.

To prove that $V(r,x)$ satisfies boundary condition (4) for each fixed x where $f(x)$ is continuous we first note that, in view of Legendre's representation (9), series (8) converges to $f(x)$ when $r = c$. But the sequence of functions $(r/c)^n$, $n = 0, 1, 2, \ldots$ is bounded, and monotonic with respect to n. Hence Abel's test (Chap. 10) shows that series (8) is uniformly convergent with respect to r $(0 \leqq r \leqq c)$. Therefore V is a continuous function of r, and $V(c - 0,x) = V(c,x) = f(x)$.

The terms in series (8) can be written as the product of the three factors A_n/n, $P_n(x)$, and $n(r/c)^n$ when $n > 0$. The first two factors are bounded for all n and x involved here. The third is nonnegative and not greater than $n(r_0/c)^n$ if $0 \leqq r \leqq r_0$. When $r_0 < c$, the series with terms $n(r_0/c)^n$ converges. Therefore series (8) converges uniformly with respect to r and x when

$0 \leq r \leq r_0$ and $-1 \leq x \leq 1$.   Thus V is continuous everywhere interior to the sphere and bounded when $0 \leq r \leq r_0 < c$.

But the series with terms $n^k(r_0/c)^n$ also converges whenever $0 < r_0 < c$, for each fixed positive value of k.   Since $n^{-2}P_n'(x)$ and $n^{-4}P_n''(x)$ are bounded $(-1 \leq x \leq 1, n = 1, 2, \ldots)$, according to Theorem 1, it follows readily that series (8) is twice differentiable with respect to r and x when $r < c$ and that the derivatives of V are continuous interior to the sphere.   Since each term of the series satisfies Laplace's equation, the sum of the series satisfies that equation.

This completes the verification of formula (8).   The solution (12) for the harmonic function W in the external region can be established in the same manner.   By writing s/c for c/r in our formula for W we find from the above discussion of series (8) that rW is bounded for large values of r $(s \leq c_0 < c)$ and for all x $(-1 \leq x \leq 1)$.

90. Steady Temperatures in a Hemisphere.   The base $\theta = \frac{1}{2}\pi$ of a solid hemisphere $r \leq 1, 0 \leq \theta \leq \frac{1}{2}\pi$ is insulated.   The flux

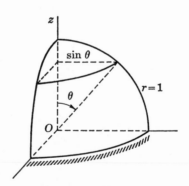

FIG. 25

of heat through the hemispherical surface is kept at prescribed values, $f(\cos \theta)$, such that the resultant rate of flow per unit time through that surface is zero in order that temperatures can be steady.   That is (Fig. 25), f satisfies the condition

$$\int_0^{\pi/2} f(\cos \theta)2\pi \sin \theta \, d\theta = 0$$

or, writing $x = \cos \theta$,

(1) $$\int_0^1 f(x) \, dx = 0.$$

If u denotes temperatures as a function of r and θ, or r and x, the condition that the base is insulated becomes

$$\frac{1}{r}\frac{\partial u}{\partial \theta} = 0, \qquad\qquad \text{when } \theta = \tfrac{1}{2}\pi.$$

But $u_\theta = -u_x \sin \theta$, so that condition on $u(r,x)$ is

$$(2) \qquad\qquad u_x(r,0) = 0 \qquad\qquad (0 < r < 1).$$

The boundary value problem in $u(r,x)$ consists of Laplace's equation

$$(3) \quad r(ru)_{rr} + [(1 - x^2)u_x]_x = 0 \qquad (0 < r < 1, 0 < x < 1),$$

condition (2), and the flux condition

$$(4) \qquad\qquad Ku_r(1,x) = f(x) \qquad\qquad (0 < x < 1),$$

where K is thermal conductivity. We assume that f and f' are sectionally continuous and that f satisfies condition (1). Also u is to satisfy the usual continuity conditions when $r < 1$ and $0 \leqq x \leqq 1$.

Separating variables by replacing u in equations (2) and (3) by $R(r)X(x)$, we obtain the conditions

$$(5) \qquad \begin{aligned} [(1 - x^2)X']' + \lambda X &= 0 \qquad (0 < x < 1), \\ X'(0) &= 0, \end{aligned}$$

where X and X' are to be continuous when $0 \leqq x \leqq 1$, and the condition $r(rR)'' = \lambda R$, where R is to be continuous. The singular Sturm-Liouville problem (5) has eigenvalues $\lambda = 2n(2n + 1)$ and eigenfunctions $X = P_{2n}(x)$ (Sec. 83). We now find that $R = r^{2n}$. Formally, then,

$$u = \sum_{n=0}^{\infty} B_n r^{2n} P_{2n}(x)$$

if B_n are such that condition (4) is satisfied; that is, if

$$(6) \qquad\qquad 2K \sum_{n=1}^{\infty} nB_n P_{2n}(x) = f(x) \qquad (0 < x < 1).$$

This is the representation of f on the interval $(0,1)$ in series of Legendre polynomials of even degree (Sec. 87) if $2nKB_n = A_{2n}$, where

$$(7) \qquad A_{2n} = (4n + 1) \int_0^1 f(\xi)P_{2n}(\xi)\,d\xi \qquad (n = 1, 2, \ldots)$$

and if f is such that $A_0 = 0$, which is precisely condition (1). Thus B_0 is left arbitrary and

$$(8) \quad u = B_0 + \frac{1}{2K} \sum_{n=1}^{\infty} \frac{1}{n} A_{2n} r^{2n} P_{2n}(x) \qquad (r \leqq 1, 0 \leqq x \leqq 1),$$

where the coefficients A_{2n} have the values (7).

The constant B_0 is the temperature at the origin $r = 0$. Solutions of Neumann problems are determined up to such an arbitrary, added constant because all boundary conditions prescribe values of derivatives of the function.

91. Other Orthogonal Sets. Laplace's equation for a function V of all three spherical coordinates r, ϕ, θ, where we again write $x = \cos \theta$, is

$$(1) \qquad r(rV)_{rr} + [(1 - x^2)V_x]_x + (1 - x^2)^{-1}V_{\phi\phi} = 0.$$

If V and its derivatives are to be continuous throughout the interior $r < c$ of a sphere, those functions must be periodic in ϕ, with period 2π. Then separation of variables leads to the particular solutions

$$(2) \quad r^n(a_m \cos m\phi + b_m \sin m\phi)P_n{}^m(x) \qquad (m,n = 0, 1, 2, \ldots).$$

The functions $X = P_n{}^m(x)$, called *associated Legendre functions*, satisfy the differential equation

$$(3) \qquad [(1 - x^2)X']' + [n(n + 1) - m^2(1 - x^2)^{-1}]X = 0.$$

Those generalizations of P_n are related to the Legendre polynomials as follows:

$$(4) \quad P_n{}^m(x) = (1 - x^2)^{\frac{1}{2}m} \frac{d^m}{dx^m} P_n(x) \qquad (m,n = 0, 1, 2, \ldots).$$

For each fixed m the set $P_n{}^m(x)$ $(n = 0, 1, 2, \ldots)$ is orthogonal on the interval $(-1,1)$ with weight function unity.[1] Note that $P_n{}^0(x) = P_n(x)$, and $P_n{}^m(x) = 0$ if $m > n$.

Other generalizations of Legendre polynomials include the *Jacobi polynomials* $P_n{}^{(\alpha,\beta)}(x)$, where $n = 0, 1, 2, \ldots, \alpha > -1$

[1] For treatments of associated Legendre functions see chap. 15 of Whittaker and Watson, "Modern Analysis," or chap. 3 of Hobson, "Spherical and Ellipsoidal Harmonics." An application is given in sec. 9.11 of Carslaw and Jaeger, "Conduction of Heat in Solids."

and $\beta > -1$, defined as

$$\frac{(-1)^n}{2^n n!} (1 - x)^{-\alpha}(1 + x)^{-\beta} \frac{d^n}{dx^n} [(1 - x)^{\alpha+n}(1 + x)^{\beta+n}].$$

When α and β are fixed, those polynomials are orthogonal on the interval $(-1,1)$ with weight function $(1 - x)^\alpha(1 + x)^\beta$. When $\alpha = \beta = 0$, they reduce to $P_n(x)$, according to Rodrigues' formula.[1]

PROBLEMS

1. If V is harmonic throughout the domains $r < c$ and $r > c$, if $V \to 0$ as $r \to \infty$, and $V = 1$ on the spherical surface $r = c$, show from the results found in Sec. 89 that $V = 1$ when $r \leqq c$ and $V = c/r$ when $r \geqq c$.

2. Write a formula for the steady temperatures $u(r,\theta)$ in a solid sphere $r \leqq 1$ if, for all ϕ, $u(1,\theta) = 1$ when $0 < \theta < \frac{1}{2}\pi$ and $u(1,\theta) = 0$ when $\frac{1}{2}\pi < \theta < \pi$.

$$Ans. \quad u = \frac{1}{2} + \frac{1}{2} \sum_{n=0}^{\infty} [P_{2n}(0) - P_{2n+2}(0)]r^{2n+1}P_{2n+1}(\cos \theta).$$

3. The base $\theta = \frac{1}{2}\pi$, $r < 1$ of a solid hemisphere $r \leqq 1$, $0 \leqq \theta \leqq \frac{1}{2}\pi$ is kept at temperature $u = 0$, while $u = 1$ on the hemispherical surface $r = 1$, $0 < \theta < \frac{1}{2}\pi$. Derive the formula

$$u(r,\theta) = \sum_{n=0}^{\infty} (-1)^n \frac{4n + 3}{2n + 2} \frac{(2n)!}{2^{2n}(n!)^2} r^{2n+1}P_{2n+1}(\cos \theta)$$

for the steady temperatures in the solid.

4. The base $\theta = \frac{1}{2}\pi$, $r < c$ of a solid hemisphere $r \leqq c$, $0 \leqq \theta \leqq \frac{1}{2}\pi$ is insulated. The temperature distribution on the hemispherical surface is $f(\cos \theta)$. Derive the formula

$$u(r,\theta) = \sum_{n=0}^{\infty} (4n + 1) \left(\frac{r}{c}\right)^{2n} P_{2n}(\cos \theta) \int_0^1 f(\xi)P_{2n}(\xi) \, d\xi$$

for the steady temperatures in the solid. Also, note that $u(r,\theta) = 1$ in case $f(\cos \theta) = 1$.

[1] An introduction to Jacobi polynomials, also to Hermite and Laguerre polynomials orthogonal on unbounded intervals, is given in the little book by D. Jackson on "Fourier Series and Orthogonal Polynomials." Differential equations satisfied by those polynomials are given there. Also see E. D. Rainville, "Special Functions."

5. A function V is harmonic and bounded in the unbounded domain $r > c$, $0 \leqq \phi \leqq 2\pi$, $0 \leqq \theta < \frac{1}{2}\pi$. If $V = 0$ everywhere on the flat boundary $\theta = \frac{1}{2}\pi$, $r > c$ and $V = f(\cos \theta)$ on the hemispherical boundary $r = c$, $0 < \theta < \frac{1}{2}\pi$, derive the formula

$$V(r,\theta) = \sum_{n=0}^{\infty} (4n + 3) \left(\frac{c}{r}\right)^{2n+2} P_{2n+1}(\cos \theta) \int_0^1 f(\xi) P_{2n+1}(\xi) \, d\xi.$$

6. The flux of heat $K u_r(1,\theta)$ into a solid sphere at its surface $r = 1$ is a prescribed function $f(\cos \theta)$, where f is such that the net time rate of flow of heat into the solid is zero. Thus show that $\int_{-1}^{1} f(\xi) \, d\xi = 0$. If $u = 0$ at the center $r = 0$, derive this formula for the steady temperatures $u(r,\theta)$ in the sphere $0 \leqq r \leqq 1$:

$$u = \frac{1}{2K} \sum_{n=1}^{\infty} \frac{2n + 1}{n} r^n P_n(\cos \theta) \int_{-1}^{1} f(\xi) P_n(\xi) \, d\xi.$$

7. Let $u(r,\theta)$ denote steady temperatures in a hollow sphere $a \leqq r \leqq b$ when $u(a,\theta) = f(\cos \theta)$ and $u(b,\theta) = 0$, $0 < \theta < \pi$. Derive the formula

$$u(r,\theta) = \sum_{n=0}^{\infty} A_n \frac{b^{2n+1} - r^{2n+1}}{b^{2n+1} - a^{2n+1}} \left(\frac{a}{r}\right)^{n+1} P_n(\cos \theta),$$

where
$$A_n = \frac{2n + 1}{2} \int_{-1}^{1} f(\xi) P_n(\xi) \, d\xi.$$

8. If $u(x,t)$ represents temperatures in a nonhomogeneous insulated bar $-1 \leqq x \leqq 1$ along the x axis, in which the thermal conductivity is proportional to $1 - x^2$, the heat equation takes the form

$$\frac{\partial u}{\partial t} = b \frac{\partial}{\partial x} \left[(1 - x^2) \frac{\partial u}{\partial x} \right]$$

where $b > 0$ and constant if the thermal coefficient c_0 is constant (Sec. 6 and Problem 8, Sec. 7). The ends $x = \pm 1$ are insulated because the conductivity vanishes there. If $u(x,0) = f(x)$ $(-1 < x < 1)$, derive the formula

$$u = \sum_{n=0}^{\infty} \frac{2n + 1}{2} \exp\left[-n(n + 1)bt\right] P_n(x) \int_{-1}^{1} f(\xi) P_n(\xi) \, d\xi.$$

9. When $f(x) = x^2$ $(-1 < x < 1)$ in Problem 8, show that

$$u(x,t) = \tfrac{1}{3} + (x^2 - \tfrac{1}{3}) \exp(-6bt).$$

10. The initial temperature distribution in a solid sphere $r \leqq 1$ is a prescribed function $f(\cos \theta)$ independent of the spherical coordinates r and ϕ.   If $k = 1$ and $x = \cos \theta$, $0 \leqq \theta \leqq \pi$, the temperature function $u(r,x,t)$ satisfies the heat equation in the form

$$r^2 u_t = r(ru)_{rr} + [(1 - x^2)u_x]_x$$

and the usual conditions of continuity.   Given that $u = 0$ when $r = 1$, separate variables in the problem to obtain solutions $R(r)X(x)T(t)$ of the homogeneous conditions, showing that X satisfies Legendre's equation, where $\lambda = n(n + 1)$, and that

$$r^2 R'' + 2rR' + [\mu^2 r^2 - n(n + 1)]R = 0, \qquad R(1) = 0.$$

Verify that $R = r^{-\frac{1}{2}} J_{n+\frac{1}{2}}(\mu_j r)$, where $\mu = \mu_j$, are roots of the equation $J_{n+\frac{1}{2}}(\mu) = 0$.   Thus, without completing the solution for u, show that

$$R(r)X(x)T(t) = r^{-\frac{1}{2}} J_{n+\frac{1}{2}}(\mu_j r) P_n(x) \exp\left(-\mu_j^2 t\right).$$

CHAPTER 10

UNIQUENESS OF SOLUTIONS

We shall prove a few theorems with which formal solutions of boundary value problems of certain types can be established either as a solution or as the only possible solution. A multiplicity of solutions may actually arise when the statement of the problem does not demand adequate continuity of the unknown function or its derivatives. This was illustrated in Problem 10, Sec. 67.

Abel's test for uniform convergence (Theorem 1) will enable us to determine further continuity properties of solutions obtained in the form of series, properties that are useful both in establishing solutions and in proving a solution unique. The remaining theorems give conditions under which not more than one solution exists. They apply to specific types of problems, and their applications are further limited because they require a rather high degree of regularity of the functions involved.

92. Cauchy Criterion for Uniform Convergence. Let $s_n(x)$ denote the sum of the first n terms of a series of functions $X_i(x)$ which converges to the sum $s(x)$:

$$(1) \qquad s_n(x) = \sum_{i=1}^{n} X_i(x), \qquad s(x) = \lim_{n \to \infty} s_n(x).$$

Suppose the series converges uniformly with respect to x for all x on some interval. Then, as noted in Sec. 14, for each positive number ϵ there exists a number n_ϵ, independent of x, such that

$$|s(x) - s_n(x)| < \tfrac{1}{2}\epsilon \qquad \text{whenever } n > n_\epsilon$$

for every x on the interval. Let j denote any positive integer. Then

$$|s_n - s_{n+j}| = |s_n - s + s - s_{n+j}| \leqq |s - s_n| + |s - s_{n+j}| < \epsilon.$$

Thus *a necessary condition for uniform convergence of the series is that, for all positive integers j,*

$$(2) \qquad |s_{n+j}(x) - s_n(x)| < \epsilon \qquad \text{whenever } n > n_\epsilon.$$

Condition (2) is a sufficient condition, the Cauchy criterion, for convergence of the series for each fixed x even if n_ϵ were not independent of x. Hence it implies that the sum $s(x)$ exists. Thus for any fixed n and x, and for the given number ϵ, an integer $j_\epsilon(x)$ exists such that

$$(3) \qquad |s(x) - s_{n+j}(x)| < \epsilon \qquad \text{whenever } j > j_\epsilon(x).$$

To show that *condition (2) implies uniform convergence,* let n denote a particular integer greater than n_ϵ, where n_ϵ corresponds to the given number ϵ in the sense that condition (2) is satisfied for all x. Then for each particular x condition (3) is satisfied when $j > j_\epsilon(x)$ and, since

$$|s - s_n| = |s - s_{n+j} + s_{n+j} - s_n| \leqq |s - s_{n+j}| + |s_{n+j} - s_n|,$$

it follows from conditions (3) and (2) that

$$(4) \qquad |s(x) - s_n(x)| < 2\epsilon$$

when $j > j_\epsilon(x)$ and $n > n_\epsilon$. Thus $|s(x) - s_n(x)|$, which is independent of j, is arbitrarily small for each x when $n > n_\epsilon$. Since n_ϵ is independent of x, the uniform convergence is established.

Note that x here may equally well denote a set of independent variables $(x_1, x_2, \ldots, x_m)$ representing all points in some set E_m of points in m dimensional space. The uniform convergence is then with respect to all m variables in E_m.

93. Abel's Test for Uniform Convergence. We now establish a test for uniform convergence of infinite series whose terms are products of functions of certain classes. Applications of the test were made earlier (Secs. 61, 79, and 89) for the purpose of verifying formal solutions of boundary value problems.

The functions in a sequence of functions $T_i(t)$, $i = 1, 2, \ldots$ are *uniformly bounded* for all points t on an interval if a constant K, independent of i, exists such that

$$(1) \qquad |T_i(t)| < K \qquad (i = 1, 2, \ldots)$$

for all t on the interval.   The sequence is *monotonic with respect to i* if for every t on the interval either

$$(2) \qquad\qquad T_{i+1}(t) \leqq T_i(t) \qquad\qquad (i = 1, 2, \ldots)$$

or else, for every t,

$$(3) \qquad\qquad T_{i+1}(t) \geqq T_i(t) \qquad\qquad (i = 1, 2, \ldots)$$

The following generalized form of a test due to Abel shows that when the terms of a uniformly convergent series are multiplied by functions $T_i(t)$ of the type just described, the new series is uniformly convergent.

Theorem 1.   *The series*

$$(4) \qquad\qquad \sum_{i=1}^{\infty} X_i(x) T_i(t)$$

converges uniformly with respect to the two variables x and t together in a region R of the xt plane provided that (a) the series $\displaystyle\sum_{i=1}^{\infty} X_i(x)$ converges uniformly with respect to x for all x such that (x,t) is in R, and (b) for all t such that (x,t) is in R, the functions $T_i(t)$ are uniformly bounded and monotonic with respect to i $(i = 1, 2, \ldots)$.

Let S_n denote the partial sum of series (4):

$$S_n(x,t) = \sum_{i=1}^{n} X_i(x) T_i(t).$$

As shown in the preceding section, the uniform convergence of that series will be established if we prove that to each positive number ϵ there corresponds an integer n_ϵ, independent of x and t, such that

$$|S_m(x,t) - S_n(x,t)| < \epsilon \qquad \text{whenever } n > n_\epsilon$$

for all integers $m = n + 1, n + 2, \ldots$ and for all points (x,t) in R.

We write s_n for the partial sum

$$s_n(x) = X_1(x) + X_2(x) + \cdots + X_n(x).$$

Then for each pair of integers m, n $(m > n)$, $S_m - S_n$ can be written

$$X_{n+1}T_{n+1} + X_{n+2}T_{n+2} + \cdots + X_m T_m$$
$$= (s_{n+1} - s_n)T_{n+1} + (s_{n+2} - s_{n+1})T_{n+2} + \cdots + (s_m - s_{m-1})T_m$$
$$= (s_{n+1} - s_n)T_{n+1} + (s_{n+2} - s_n)T_{n+2} - (s_{n+1} - s_n)T_{n+2}$$
$$+ \cdots + (s_m - s_n)T_m - (s_{m-1} - s_n)T_m.$$

By pairing alternate terms here we find that

$$(5) \quad S_m - S_n = (s_{n+1} - s_n)(T_{n+1} - T_{n+2}) + (s_{n+2} - s_n)(T_{n+2} - T_{n+3})$$
$$+ \cdots + (s_{m-1} - s_n)(T_{m-1} - T_m) + (s_m - s_n)T_m.$$

Suppose now that the functions T_i are nonincreasing with respect to i so that they satisfy condition (2). They also satisfy the uniform boundedness condition (1). Then the factors $T_{n+1} - T_{n+2}$, $T_{n+2} - T_{n+3}$, etc., in equation (5) are nonnegative and $|T_i(t)| < K$. Since the series with terms $X_i(x)$ converges uniformly, an integer n_ϵ exists such that

$$|s_{n+j}(x) - s_n(x)| < \frac{\epsilon}{3K} \qquad \text{whenever } n > n_\epsilon$$

for all positive integers j, where ϵ is any given positive number and n_ϵ is independent of x (Sec. 92). Then if $n > n_\epsilon$ and $m > n$, it follows from equation (5) that

$$|S_m - S_n| < \frac{\epsilon}{3K}[(T_{n+1} - T_{n+2}) + (T_{n+2} - T_{n+3}) + \cdots + |T_m|]$$

$$= \frac{\epsilon}{3K}(T_{n+1} - T_m + |T_m|) \leq \frac{\epsilon}{3K}(|T_{n+1}| + 2|T_m|).$$

Therefore $|S_m(x,t) - S_n(x,t)| < \epsilon$ whenever $m > n > n_\epsilon$,

and the uniform convergence of series (4) is established.

The proof is similar in case the functions T_i are nondecreasing with respect to i.

When x is kept fixed, the series with terms X_i is a series of constants, and the only requirement placed on that series is that it be convergent. Then the theorem shows that, when T_i are bounded and monotonic, the series of terms $X_i T_i(t)$ is uniformly convergent with respect to t.

Extensions of the theorem to cases in which X_i are functions of x and t, or where X_i and T_i are functions of several variables, become evident when it is observed that our proof rests on the uniform convergence of the series of terms X_i and the bounded monotonic nature of the functions T_i.

94. Uniqueness of Solutions of the Heat Equation.   Let D denote the domain consisting of all points interior to a closed surface S, and let $\bar{D}$ be the closure of that domain, consisting of all points in D and all points on S.   We assume always that the closed surface S is *piecewise smooth;* that is, it is a continuous surface consisting of a finite number of parts over each of which the outward-drawn unit normal vector varies continuously from point to point.   Then if W is a function of x, y, and z continuous in $\bar{D}$, together with its partial derivatives of first and second order, an extended form of the divergence theorem states that

$$(1) \qquad \iint_S W \frac{dW}{dn}\, dA = \iiint_D (W\nabla^2 W + W_x{}^2 + W_y{}^2 + W_z{}^2)\, dV.$$

Here dA is the area element on S, dV represents $dx\, dy\, dz$, and dW/dn is the derivative in the direction of the outward-drawn normal to S.[1]

Consider a homogeneous solid whose interior is the domain D and whose temperatures at time t are denoted by $u(x, y, z, t)$. A fairly general problem in heat conduction is the following.

$$(2) \qquad\qquad u_t = k\nabla^2 u + \phi(x,y,z,t) \qquad [(x,y,z) \text{ in } D,\, t > 0],$$
$$(3) \qquad\qquad u(x,y,z,0) = f(x,y,z) \qquad\qquad [(x,y,z) \text{ in } \bar{D}],$$
$$(4) \qquad\qquad u = g(x,y,z,t) \qquad\qquad [(x,y,z) \text{ on } S,\, t \geqq 0].$$

This is the problem of determining temperatures in a body with prescribed initial temperatures $f(x,y,z)$ and surface temperatures $g(x,y,z,t)$, interior to which heat may be generated continuously at a rate per unit volume per unit time proportional to $\phi(x,y,z,t)$.

Suppose the problem has two solutions

$$u = u_1(x,y,z,t), \qquad u = u_2(x,y,z,t),$$

where both u_1 and u_2 are continuous functions in the closed region $\bar{D}$ when $t \geqq 0$, while their derivatives of first order with respect to t and of first and second order with respect to x, y, and z are

---

[1] Formula (1) is found by applying the basic divergence theorem used in Sec. 6 to the vector field W grad W.   See, for instance, W. Kaplan, "Advanced Calculus," p. 274, 1952.

continuous in $\bar{D}$ when $t > 0$.   Since u_1 and u_2 satisfy the linear conditions (2), (3), and (4) their difference

$$w(x,y,z,t) = u_1(x,y,z,t) - u_2(x,y,z,t)$$

satisfies the homogeneous problem

(5) $$w_t = k\nabla^2 w \qquad [(x,y,z) \text{ in } D, \ t > 0],$$
(6) $$w(x,y,z,0) = 0 \qquad [(x,y,z) \text{ in } \bar{D}],$$
(7) $$w = 0 \qquad [(x,y,z) \text{ on } S, \ t \geqq 0].$$

Moreover, w and its derivatives have the continuity properties of u_1 and u_2 assumed above.

We shall show now that $w = 0$ in D when $t > 0$ so that the two solutions u_1 and u_2 are identical.   It will follow that not more than one solution of the boundary value problem in u can exist if the solution is required to satisfy the continuity conditions stated.

The continuity of w with respect to x, y, z, and t together in the closed region $\bar{D}$ when $t \geqq 0$ implies that the integral

(8) $$I(t) = \tfrac{1}{2} \iiint_D [w(x,y,z,t)]^2 \, dV$$

is a continuous function of t when $t \geqq 0$.   According to condition (6), $I(0) = 0$.   In view of the continuity of w_t when $t > 0$ and equation (5), we can write

$$I'(t) = \iiint_D w w_t \, dV = k \iiint_D w \nabla^2 w \, dV \qquad (t > 0).$$

The divergence formula (1) applies to the last integral because of the continuity of the derivatives of w when $t > 0$; thus

(9) $$\iiint_D w \nabla^2 w \, dV = \iint_S w \frac{dw}{dn} \, dA - \iiint_D (w_x{}^2 + w_y{}^2 + w_z{}^2) \, dV$$

when $t > 0$.   But $w = 0$ on S; therefore

$$I'(t) = -k \iiint_D (w_x{}^2 + w_y{}^2 + w_z{}^2) \, dV \leqq 0.$$

The mean value theorem applies to $I(t)$.   Thus for each positive t a number t_1 $(0 < t_1 < t)$ exists such that

$$I(t) - I(0) = t I'(t_1),$$

and since $I(0) = 0$ and $I'(t_1) \leq 0$, it follows that $I(t) \leq 0$. However definition (8) of the integral shows that $I(t) \geq 0$. Therefore

$$I(t) = 0 \qquad (t \geq 0),$$

and so the nonnegative integrand w^2 cannot have a positive value at any point in D, for if so the continuity of w^2 requires that w^2 be positive throughout a neighborhood of the point and then $I(t) > 0$. Consequently

$$w(z,y,z,t) = 0 \qquad [(x,y,z) \text{ in } \bar{D},\ t \geq 0],$$

and the following theorem on uniqueness is established.

Theorem 2. *Let u satisfy these conditions of regularity: (a) it is a continuous function of the variables x, y, z, and t, taken together, when the point (x,y,z) is in the closed region $\bar{D}$ and $t \geq 0$; (b) those derivatives of u present in the heat equation (2) are continuous in the same sense when $t > 0$. Then if u is a solution of the boundary value problem (2) to (4), it is the only possible solution satisfying conditions (a) and (b).*

The condition that u be continuous in $\bar{D}$ when $t = 0$ restricts the usefulness of our theorem. It is clearly not satisfied if the initial temperature function f in condition (3) fails to be continuous throughout $\bar{D}$, or if at some point on S the initial value $g(x,y,z,0)$ of the prescribed surface temperature differs from the value $f(x,y,z)$. The continuity requirement at $t = 0$ can be relaxed in some cases.[1]

When conditions (a) and (b) are added to the requirement that u is to satisfy the heat equation and the boundary conditions, our boundary value problem is completely stated, provided it has a solution, for that will be the only possible solution.

The proof of Theorem 2 required that the integral

$$\iint_S w \frac{dw}{dn}\, dA$$

in formula (9) should either vanish or have a negative value. It vanished because $w = 0$ on S. But it is never positive if

---

[1] See, for example, sec. 73 of the author's "Operational Mathematics," 1958.

condition (4) is replaced by the boundary condition

$$(10) \qquad \frac{du}{dn} + hu = g(x,y,z,t) \qquad [(x,y,z) \text{ on } S, \, t > 0]$$

where $h \geq 0$. For in that case $dw/dn = -hw$ on S, and $w \, dw/dn \leq 0$. Thus our theorem can be modified as follows.

Theorem 3. *The statement in Theorem 2 is true if boundary condition* (4) *is replaced by condition* (10), *or if* (4) *is satisfied on part of the surface S and* (10) *on the rest.*

95. Example. In the problem of temperature distribution in a slab with insulated faces $x = 0$ and $x = \pi$ and initial temperature $f(x)$ (Sec. 62b), assume f continuous and f' sectionally continuous over the interval $0 \leq x \leq \pi$. Then the Fourier cosine series for f converges uniformly to $f(x)$ on that interval.

Let $u(x,t)$ denote the sum of the series

$$(1) \qquad \tfrac{1}{2}a_0 + \sum_{n=1}^{\infty} a_n \exp(-n^2 kt) \cos nx \qquad (0 \leq x \leq \pi, \, t \geq 0),$$

obtained as a formal solution of the problem, a_n being the coefficients in the Fourier cosine series for f.

We can see from Abel's test (Theorem 1) that series (1) converges uniformly with respect to x and t together in the region $0 \leq x \leq \pi$, $t \geq 0$; thus u is continuous there. When $t \geq t_0$, where t_0 is any positive number, the series obtained by differentiating series (1) term by term, any number of times with respect to x or t, is uniformly convergent according to the Weierstrass test. It follows readily that u not only satisfies all conditions of the boundary value problem (compare Sec. 61), but also that u_t, u_x, and u_{xx} are continuous functions in the region $0 \leq x \leq \pi$, $t > 0$. Thus u satisfies the regularity conditions (a) and (b) of Theorem 2.

The temperature problem for the slab is the same as the problem for a prismatic bar with its bases in the planes $x = 0$ and $x = \pi$ and with its lateral surface, parallel to the x axis, insulated $(du/dn = 0)$. Let the domain D consist of all interior points of the prism. Then Theorem 3 applies to show that the sum $u(x,t)$ of series (1) is the only solution that satisfies the conditions (a) and (b).

PROBLEMS

1. In the problem of temperature distribution in the hemisphere $r \leqq 1, 0 \leqq \theta \leqq \frac{1}{2}\pi$, solved formally in Sec. 90, assume that f and f' are sectionally continuous. With the aid of Abel's test and Legendre's series prove that for each fixed x $(0 < x < 1, x = \cos \theta)$, where f is continuous, the series in the solution (8), Sec. 90, namely

$$B_0 + \frac{1}{2K} \sum_{n=1}^{\infty} \frac{1}{n} A_{2n} r^{2n} P_{2n}(x),$$

converges uniformly with respect to r $(0 \leqq r \leqq 1)$. Also prove that the series is differentiable with respect to r and that

$$Ku_r(1 - 0, x) = f(x),$$

where $u(r,x)$ is the sum of the series.

2. In Problem 10, Sec. 62, on temperatures $u(x,t)$ in a slab initially at temperatures $f(x)$, throughout which heat is generated at a constant rate, assume f continuous and f' sectionally continuous $(0 \leqq x \leqq \pi)$ and that $f(0) = f(\pi) = 0$. Prove that the function $u(x,t)$ given there is the only solution of the problem which satisfies the regularity conditions (a) and (b) stated in Theorem 2.

3. Establish the solution of Problem 13, Sec. 62, and prove that it is the only possible solution satisfying the regularity conditions (a) and (b) stated in Theorem 2. Note that in this case the Weierstrass test suffices for all proofs of uniform convergence.

4. In Sec. 79, given that the initial temperature function $f(\rho)$ for the cylinder is continuous $(0 \leqq \rho \leqq c)$ and such that its Fourier-Bessel series used there converges uniformly to $f(\rho)$ on that interval, prove that the solution established there is the only one that satisfies our regularity conditions.

96. Solutions of Laplace's or Poisson's Equation.

Let W be a harmonic function in a domain D of three-dimensional space bounded by a continuous closed surface S that is piecewise smooth. Assume also that W and its partial derivatives of the first order are continuous in the closure $\bar{D}$ of the domain. Then since

$$(1) \qquad \nabla^2 W(x,y,z) = 0 \qquad [(x,y,z) \text{ in } D],$$

form (1), Sec. 94, of the divergence formula becomes

$$(2) \qquad \iint_S W \frac{dW}{dn} dA = \iiint_D (W_x{}^2 + W_y{}^2 + W_z{}^2) dV.$$

This formula is valid for our function W even though we have required the derivatives of second order to be continuous only in D, not in the closed region $\bar{D}$. It is not difficult to prove that, because $\nabla^2 W = 0$, modification of the usual conditions in the divergence theorem is possible.[1]

Suppose that $W = 0$ at all points of S. Then the first integral in formula (2), and therefore the second, vanishes. But the integrand of the second is nonnegative and continuous in $\bar{D}$. It must therefore vanish throughout $\bar{D}$; that is,

$$(3) \qquad\qquad W_x = W_y = W_z = 0 \qquad\qquad [(x,y,z) \text{ in } \bar{D}].$$

Consequently $W(x,y,z)$ is constant; but it is zero on S and continuous in $\bar{D}$, and therefore $W = 0$ throughout $\bar{D}$.

Suppose that dW/dn, instead of W, vanishes on S; or to make the condition more general, suppose

$$(4) \qquad\qquad \frac{dW}{dn} + hW = 0 \qquad\qquad [(x,y,x) \text{ on } S],$$

where $h \geqq 0$ and h is either a constant or a function of x, y, and z. Then on S

$$W \frac{dW}{dn} = -hW^2 \leqq 0$$

so that the first integral in formula (2) is nonpositive. But the second integral is nonnegative; hence it must vanish and again conditions (3) follow so that W is constant in $\bar{D}$.

In case W vanishes over part of S and satisfies condition (4) on the rest of that surface, our argument still shows that W is constant in $\bar{D}$. In this case the constant must be zero.

Now let U denote a function continuous in $\bar{D}$, together with its partial derivatives of the first order, with continuous derivatives of the second order in D, which satisfies these conditions:

$$(5) \qquad\qquad \nabla^2 U(x,y,z) = f(x,y,z) \qquad\qquad [(x,y,z) \text{ in } D],$$

$$(6) \qquad\qquad p \frac{dU}{dn} + hU = g \qquad\qquad [(x,y,z) \text{ on } S].$$

Here f, p, h, and g denote prescribed constants or functions of (x,y,z); but we assume that $p \geqq 0$ and $h \geqq 0$.

Boundary condition (6) includes important special cases. When $p = 0$ on S, or on part of S, the value of U is assigned

[1] See O. D. Kellogg, "Foundations of Potential Theory," p. 119.

there.   When $h = 0$, the value of dU/dn is assigned.   Of course p and h must not vanish simultaneously.

If $U = U_1(x,y,z)$ and $U = U_2(x,y,z)$ are two solutions of this problem, their difference

$$W = U_1 - U_2$$

satisfies Laplace's equation in D and the condition

$$p\frac{dW}{dn} + hW = 0$$

on S.   Moreover W satisfies the conditions of regularity required of U_1 and U_2.   Thus it is harmonic in D, and W and its derivatives of first order are continuous in $\bar{D}$.   It follows from the results established above for harmonic functions that W must be constant throughout $\bar{D}$.   Thus $dW/dn = 0$ on S.   In case $h \neq 0$ at some point of S, then W vanishes there so that $W = 0$ throughout $\bar{D}$.

We have now established the following uniqueness theorem for problems in electrostatic or gravitational potential, steady temperatures, or other boundary value problems for Laplace's or Poisson's equation.

Theorem 4.   *Let $U(x,y,z)$ satisfy these conditions of regularity in a domain D bounded by a closed surface S: (a) it is continuous, together with its derivatives of first order, in $\bar{D}$, and (b) its derivatives of second order are continuous in D.   Then if U is a solution of the boundary value problem consisting of conditions (5) and (6), it is the only solution satisfying conditions (a) and (b), except possibly for $U + C$, where C is an arbitrary constant.   Unless $h = 0$ at every point of S, then $C = 0$, and the solution is unique.*

It is possible to show that this theorem also applies when D is the unbounded domain exterior to the closed surface S, provided U satisfies the additional requirement that the absolute values of rU, r^2U_{xx}, r^2U_{yy}, and r^2U_{zz} shall be bounded for all r greater than some fixed number, where r is the distance from (x,y,z) to the origin.[1]   Then since U vanishes as $r \to \infty$, the constant C is zero and the solution is unique.   But note that S is a closed surface, so that this extension of the theorem does not apply, for instance, to unbounded domains between two planes or inside a cylinder.

[1] Kellogg, *op. cit.*

Condition (a) in Theorem 4 is severe because it requires U and its derivatives of first order to be continuous on the surface S. For problems in which $p = 0$ on S, so that U is prescribed on the entire boundary, the condition can be relaxed so as to require only the continuity of U itself in $\bar{D}$, if derivatives are continuous in D. This follows directly from a fundamental theorem in potential theory: if a function other than a constant is harmonic in D and continuous in $\bar{D}$, its maximum and minimum values are assumed at points on S, never in D.[1]

97. An Application. To illustrate the use of Theorem 4 consider the problem in Sec. 63 of determining steady temperatures $u(x,y)$ in a rectangular plate with three edges at temperature zero and an assigned temperature distribution on the fourth edge. The faces of the plate are insulated. For our illustration it is sufficient to consider a square plate with edge length π. As long as $du/dn = 0$ on the faces, the thickness of the plate does not affect the problem.

The domain D is the interior of the region bounded by the planes $x = 0$, $x = \pi$, $y = 0$, $y = \pi$, $z = z_1$, and $z = z_2$, where z_1 and z_2 are constants. Then S is the boundary of that domain. The required function u is harmonic in D. It vanishes on the three parts $x = 0$, $x = \pi$, and $y = \pi$ of S; and $u = f(x)$ on the part $y = 0$; also, $u_z = 0$ on the parts $z = z_1$ and $z = z_2$. Thus Theorem 4 applies if u and its derivatives of first order are continuous in $\bar{D}$.

First, suppose u is independent of z. Then

$$(1) \qquad u_{xx}(x,y) + u_{yy}(x,y) = 0 \quad (0 < x < \pi, 0 < y < \pi),$$

$$(2) \qquad u(0,y) = u(\pi,y) = 0 \qquad\qquad (0 \leqq y \leqq \pi),$$

$$(3) \qquad u(x,0) = f(x), \qquad u(x,\pi) = 0 \qquad (0 \leqq x \leqq \pi).$$

The formal solution found in Sec. 63 becomes

$$(4) \qquad u = \sum_{n=1}^{\infty} b_m \frac{\sinh n(\pi - y)}{\sinh n\pi} \sin nx,$$

where b_n are the coefficients in the Fourier sine series for f.

[1] Physically, the theorem seems evident since it states that steady temperatures cannot have maximum or minimum values interior to a solid in which no heat is generated. For a proof in three dimensions see Kellogg, *op. cit.*; in two dimensions, see Churchill, "Complex Variables and Applications," sec. 54, 1960.

To show that the function (4) satisfies the regularity conditions, let us require that f and f' be continuous and f'' sectionally continuous $(0 \leqq x \leqq \pi)$ and that

$$f(0) = f(\pi) = 0.$$

The results found in Chap. 5 then show that

$$(5) \qquad f(x) = \sum_{n=1}^{\infty} b_n \sin nx, \qquad f'(x) = \sum_{n=1}^{\infty} nb_n \cos nx,$$

and both series converge uniformly on the interval $0 \leqq x \leqq \pi$. The second series, obtained by differentiating the first, is the Fourier cosine series for f' and, since f' is continuous and f'' sectionally continuous, that series is not only uniformly convergent but the series of absolute values $|nb_n|$ of its coefficients converges.   It follows from the Weierstrass test that the series

$$(6) \qquad \sum_{n=1}^{\infty} nb_n \sin nx \qquad\qquad (0 \leqq x \leqq \pi)$$

also converges uniformly with respect to x.

Let us show that the sequence of functions

$$(7) \qquad \frac{\sinh n(\pi - y)}{\sinh n\pi} \quad (n = 1, 2, \ldots, 0 \leqq y \leqq \pi)$$

appearing in series (4) is monotonic and nonincreasing as n increases, for each fixed y.   This is evident when $y = 0$ and when $y = \pi$.   It is true when $0 < y < \pi$ if the function

$$T(t) = \frac{\sinh \beta t}{\sinh \alpha t} \qquad (t > 0, \alpha > \beta > 0)$$

always decreases as t grows.   Now

$$T'(t) \sinh^2 \alpha t = \beta \sinh \alpha t \cosh \beta t - \alpha \sinh \beta t \cosh \alpha t$$
$$= -\tfrac{1}{2}(\alpha - \beta) \sinh (\alpha + \beta)t + \tfrac{1}{2}(\alpha + \beta) \sinh (\alpha - \beta)t$$
$$= -\frac{\alpha^2 - \beta^2}{2} \left[\frac{\sinh (\alpha + \beta)t}{\alpha + \beta} - \frac{\sinh (\alpha - \beta)t}{\alpha - \beta} \right]$$
$$= -\frac{\alpha^2 - \beta^2}{2} \sum_{n=0}^{\infty} \frac{(\alpha + \beta)^{2n} - (\alpha - \beta)^{2n}}{(2n + 1)!} t^{2n+1}.$$

The terms of this series are positive, so that $T'(t) < 0$; therefore $T(t)$ decreases as t grows.

Likewise the positive-valued functions

$$(8) \qquad \frac{\cosh n(\pi - y)}{\sinh n\pi} \qquad (0 \leqq y \leqq \pi),$$

arising in the series for u_y, never increase in value as n grows because their squares can be written

$$(9) \qquad \frac{1}{\sinh^2 n\pi} + \left[\frac{\sinh n(\pi - y)^2}{\sinh n\pi} \right],$$

and each term is nonincreasing.

The values of functions (7) clearly vary only from zero to unity for all n and y involved. Functions (8) are also uniformly bounded. Therefore those functions can be used in Abel's test for uniform convergence. From the uniform convergence of the series in equations (5), and series (6), on the interval $0 \leqq x \leqq \pi$, we conclude not only that series (4) converges uniformly with respect to x and y together in the region $0 \leqq x \leqq \pi$, $0 \leqq y \leqq \pi$ but also that the uniform convergence holds true for the series obtained by differentiating series (4) termwise, once, with respect to either x or y.

Consequently series (4) is differentiable with respect to x and y; also its sum $u(x,y)$ and u_x and u_y are continuous in the closed region $0 \leqq x \leqq \pi$, $0 \leqq y \leqq \pi$. Clearly u satisfies boundary conditions (2) and (3).

The derivatives of second order with respect to either x or y, of the terms in series (4), have absolute values not greater than

$$(10) \qquad n^2 |b_n| \frac{\sinh n(\pi - y_0)}{\sinh n\pi}$$

when $0 \leqq x \leqq \pi$ and $y_0 \leqq y \leqq \pi$, where $y_0 > 0$. Let M be chosen such that $|b_n| < M$ for all n. Then from the inequalities

$$2 \sinh n(\pi - y_0) < \exp n(\pi - y_0), \qquad 2 \sinh n\pi \geqq e^{n\pi}(1 - e^{-2\pi}),$$

it follows that the numbers (10) are less than

$$\frac{M}{1 - \exp(-2\pi)} n^2 \exp(-ny_0).$$

The series with these terms converges, according to the ratio test, since $y_0 > 0$, and so the Weierstrass test ensures the uniform

convergence of the series of second derivatives of terms in series
(4) when $y_0 \leqq y \leqq \pi$. Thus series (4) is twice differentiable;
also u_{xx} and u_{yy} are continuous in the region $0 \leqq x \leqq \pi, 0 < y \leqq \pi$.

Since the terms of series (4) satisfy Laplace's equation, the
sum $u(x,y)$ of that series satisfies condition (1). Thus u is established as a solution of our boundary value problem. Moreover
u satisfies our regularity conditions, even with respect to z, since
it is independent of z and $u_z = 0$ everywhere, at the parts $z = z_1$
and $z = z_2$ of S in particular. According to Theorem 4, the
function defined by series (4) is the only possible solution that
satisfies the regularity conditions.

98. Solutions of a Wave Equation. Consider the following
generalization of the problem solved in Sec. 56 for the transverse
displacements in a stretched string.

$$
\begin{align}
&(1) &y_{tt}(x,t) = a^2 y_{xx}(x,t) + \phi(x,t) &\qquad (0 < x < c, t > 0), \\
&(2) &y(0,t) = p(t), \qquad y(c,t) = q(t) &\qquad (t \geqq 0), \\
&(3) &y(x,0) = f(x) \qquad y_t(x,0) = g(x) &\qquad (0 \leqq x \leqq c).
\end{align}
$$

But we now require y to be of class C^2 in the region R:
$0 \leqq x \leqq c, t \geqq 0$, by which we shall mean that y and its derivatives of first and second order, including y_{xt} and y_{tx}, are to be
continuous functions in R. As indicated by examples in Chap.
7, the prescribed functions ϕ, p, q, f, and g must be restricted
if the problem is to have a solution of class C^2.

Suppose there are two solutions $y_1(x,t)$ and $y_2(x,t)$ of that class.
Then the difference $z = y_1 - y_2$ is of class C^2 in R, and it satisfies
the homogeneous problem

$$
\begin{align}
&(4) &z_{tt}(x,t) = a^2 z_{xx}(x,t) &\qquad (0 < x < c, t > 0), \\
&(5) &z(0,t) = 0, \qquad z(c,t) = 0 &\qquad (t \geqq 0), \\
&(6) &z(x,0) = 0, \qquad z_t(x,0) = 0 &\qquad (0 \leqq x \leqq c).
\end{align}
$$

We shall prove that $z = 0$ throughout R; thus $y_1 = y_2$.

The integrand of the integral

$$(7) \qquad I(t) = \tfrac{1}{2} \int_0^c (z_x{}^2 + a^{-2} z_t{}^2)\, dx \qquad\qquad (t \geqq 0)$$

satisfies conditions such that we can write

$$(8) \qquad I'(t) = \int_0^c (z_x z_{xt} + a^{-2} z_t z_{tt})\, dx.$$

Since $z_{tt} = a^2 z_{xx}$, the integrand here can be written

$$z_x z_{tx} + z_t z_{xx} = \frac{\partial}{\partial x}(z_x z_t);$$

and so, in view of conditions (5),

$$(9) \qquad I'(t) = z_x(c,t)z_t(c,t) - z_x(0,t)z_t(0,t) = 0.$$

Hence $I(t)$ is a constant. But definition (7) shows that $I(0) = 0$ because $z(x,0) = 0$ and so $z_x(x,0) = 0$; also $z_t(x,0) = 0$. Thus $I(t) = 0$. The nonnegative continuous integrand of that integral must therefore vanish; that is,

$$z_x(x,t) = z_t(x,t) = 0 \qquad (0 \leq x \leq c,\, t \geq 0),$$

so z is constant; in fact $z(x,t) = 0$ because $z(x,0) = 0$.

Thus *the boundary value problem consisting of conditions* (1), (2), *and* (3) *cannot have more than one solution of class C^2 in R.*

In case y_x, instead of y, is prescribed at the end point in either or both of conditions (2), the proof of uniqueness is still valid because condition (9) is again satisfied.

The requirement of continuity of derivatives of y is severe. Solutions of many simple problems in the wave equation have discontinuities in their derivatives.[1]

PROBLEMS

1. In the Dirichlet problem for a rectangle treated in Sec. 63, let f and f' be sectionally continuous on the interval $(0,a)$. Prove that the formal solution found there does satisfy the condition $u(x,+0) = f(x)$ when $0 < x < a$ if $f(x)$ is defined as the mean of $f(x + 0)$ and $f(x - 0)$ at its points of discontinuity.

2. Formulate a complete statement of the boundary value problem for steady temperatures in a square plate with insulated faces, when the edges $x = 0$, $x = \pi$, and $y = 0$ are insulated, and the edge $y = \pi$ is kept at temperatures $u = f(x)$. If f, f', and f'' are continuous on the interval $0 \leq x \leq \pi$ and $f'(0) = f'(\pi) = 0$, show that your problem has the

---

[1] It is possible, although rather tedious, to relax the requirement of continuity of derivatives of second order. Equation (8), where we assumed that continuity, is valid when those derivatives are permitted to have finite jumps at certain lines in the region R. In a special case uniqueness under such conditions is considered in Sec. 84 of the author's "Operational Mathematics," 1958.

unique solution

$$u = \frac{1}{2} a_0 + \sum_{n=1}^{\infty} a_n \frac{\cosh ny}{\cosh n\pi} \cos nx \quad \left[a_n = \frac{2}{\pi} \int_0^{\pi} f(x) \cos nx \, dx \right].$$

3. Replace the infinite cylinder in Sec. 67a by a cylinder bounded by the surfaces $\rho = 1$, $z = z_1$, and $z = z_2$, where $u_z = 0$ on the last two parts of the surface. Also let $f(\phi)$ be a periodic function of period 2π with a continuous derivative of second order everywhere. Then show that the function u given by formula (5), Sec. 67, is the unique solution, satisfying our conditions of regularity, of the problem in steady temperatures.

4. Use the uniqueness established in Sec. 98 to show that the solution $y = A \sin \pi x \cos \pi a t$ of Problem 1, Sec. 57, is the only solution of class C^2 in the region $0 \le x \le 1$, $t \ge 0$.

5. Show the solution of Problem 3, Sec. 60, is unique in class C^2.

6. In Sec. 56, let $f(x)$ be such that its odd periodic extension $F(s)$ has a continuous derivative F'' for all s $(-\infty < s < \infty)$; then show that the solution (9) there is unique in class C^2.

BIBLIOGRAPHY

The following list of books and articles for supplementary study of the various topics that have been introduced is far from exhaustive. Further references are given in the books listed here.

Bôcher, M.: Introduction to the Theory of Fourier's Series, *Annals of Math.*, ser. 2, vol. 7, pp. 81–152, 1906.

Bowman, F.: "Introduction to Bessel Functions," Dover Publications, New York, 1958.

Carslaw, H. S.: "Theory of Fourier's Series and Integrals," 3d ed., Macmillan & Co., Ltd., London, and Dover Publications, New York, 1930.

———, and J. C. Jaeger: "Conduction of Heat in Solids," 2d ed., Oxford University Press, London, 1959.

Churchill, R. V.: "Complex Variables and Applications," 2d ed., McGraw-Hill Book Company, Inc., New York, 1960.

———: "Operational Mathematics," 2d ed., McGraw-Hill Book Company, Inc., New York, 1958.

Coddington, E. A., and N. Levinson: "Theory of Ordinary Differential Equations," McGraw-Hill Book Company, Inc., New York, 1955.

Collatz, L.: "Eigenwertprobleme und ihre numerische Behandlung," Chelsea Publishing Company, New York, 1948.

Courant, R., and D. Hilbert: "Methods of Mathematical Physics," Interscience Publishers, Inc., New York, vol. 1, 1953; vol. 2, "Partial Differential Equations," 1962.

Epstein, B.: "Partial Differential Equations: An Introduction," McGraw-Hill Book Company, Inc., New York, 1962.

Erdélyi, A., W. Magnus, and F. Tricomi: "Higher Transcendental Functions," McGraw-Hill Book Company, Inc., New York, vols. 1,2, 1953; vol. 3, 1955.

Frank, P., and R. v. Mises: "Die Differential- und Integralgleichungen der Mechanik und Physik," vols. 1 and 2, F. Wieweg & Sohn, Braunschweig, 1930, 1955.

Friedman, B.: "Principles and Techniques of Applied Mathematics," John Wiley & Sons, Inc., New York, 1956.

Gray, A., G. B. Mathews, and T. M. MacRobert: "A Treatise on Bessel Functions and Their Applications to Physics," 2d ed., Macmillan & Co., Ltd., London, 1952.

Greenspan, D.: "Introduction to Partial Differential Equations," McGraw-Hill Book Company, Inc., New York, 1961.

Hobson, E. W.: "The Theory of Spherical and Ellipsoidal Harmonics," Cambridge University Press, London, 1931.

Ince, E. L.: "Ordinary Differential Equations," Longmans, Green & Co., Ltd., London, 1927.

Jackson, D.: "Fourier Series and Orthogonal Polynomials," Carus Mathematical Monographs, No. 6, Mathematical Association of America, 1941.

Jahnke, E., F. Emde, and F. Losch: "Tables of Higher Functions," McGraw-Hill Book Company, Inc., New York, 1960.

Kellogg, O. D.: "Foundations of Potential Theory," Springer-Verlag OHG, Berlin, 1929.

Langer, R. E.: Fourier's Series: The Genesis and Evolution of a Theory, Slaught Memorial Papers, No. 1, *Amer. Math. Monthly*, vol. 54, no. 7, part 2, pp. 1–86, 1947.

Lord Rayleigh: "Theory of Sound," 2d ed., vols, 1 and 2, Dover Publications, New York, 1945.

Rainville, E. D.: "Special Functions," The Macmillan Company, New York, 1960.

Rogosinski, W.: "Fourier Series," Chelsea Publishing Company, New York, 1950.

Sneddon, I. N.: "Fourier Transforms," McGraw-Hill Book Company, Inc., New York, 1951.

Tamarkin, J. D., and W. Feller: "Partial Differential Equations," mimeographed lecture notes, Brown University, 1941.

Titchmarsh, E. C.: "Eigenfunction Expansions Associated with Second-order Differential Equations," Oxford University Press, London, 1946, vol. 2, 1958.

————: "Theory of Fourier Integrals," Oxford University Press, London, 1937.

Tolstov, G. P.: "Fourier Series," Prentice-Hall, Inc., Englewood Cliffs, N.J., 1962.

Van Vleck, E. B.: The Influence of Fourier's Series upon the Development of Mathematics, *Science*, vol. 39, pp. 113–124, 1914.

Watson, G. N.: "A Treatise on the Theory of Bessel Functions," 2d ed., Cambridge University Press, London, 1944.

Whittaker, E. T., and G. N. Watson: "Modern Analysis," Cambridge University Press, London, 1950.

Zygmund, A.: "Trigonometric Series," 2d ed., vols. 1 and 2, Cambridge University Press, New York, 1959.

INDEX